Clinging Young: Science of In-arms Carrying

Mel Cyrille

First published in Great Britain by Melissa Cyrille in 2018.

Carried

www.melcyrille.com

www.inarmscarrying.com

The right of Melissa Cyrille to be identified as the author of this work has been asserted in accordance with the Copyright act of 1988.

Images copyright: Miriam MacMillan/Greenshoots Photography.

All rights reserved.

This book is sold subject to the condition that it shall not, by way of trade or otherwise, be lent, re-sold, hired out or otherwise circulated without the publisher's prior consent in any form of binding or cover other than that in which it is published and without similar condition including this condition being imposed on the subsequent purchaser.

No part of this publication may be reproduced or transmitted in any form without express permission from the author.

ISBN: 9781717910516

For Ulrika – my fellow in-arms geek!

Table of Contents

Acknowledgements 13

Foreword 17

Introduction 21

PART I 25

Chapter 1.1 - The Clinging Young Concept   27

Chapter 1.2 - Introduction to How Clinging Works 29
    Critical period hypothesis 29
    How a baby clings 31
    Position of legs 33
    A workable environment 35
    Endurance 36
    Variables in carrying environment 38
    Carrying is movement 40
    A passive society 42

Chapter 1.3 - The Role of Reflexes 47
    Developing the nervous system 49
    Plantar grasp and Babinski's sign 51
    Stepping reflex/Clinging adjustment reflex (CAR) 52
    Clinging reflex 55

    Upper extremity proprioceptive placing reflex and foot placing reflex 57
    Tonic Labyrinthine Reflex (TLR) 59
    Moro and startle reflexes 61
    Palmar Grasp 62
    Cremasteric reflex 63
    Sleeping 64
    Reflexive carrying in action 64
    A matter of repetition? 66
    The premature infant 69

Chapter 1.4 - Breastfeeding and Carrying 71
    Breast crawl 73
    Biological nurturing 74
    Postpartum recovery 76
    Feeding whilst carrying 78
    The babywearing connection 79

PART II 81

Chapter 2.1 - Participation in Carrying – Where Does it Begin? 83
    Communication from baby 86
    Initial contact 92
    Lifting 93
    Bodily approach 94
    Bodily contact 95

Summary 97

Chapter 2.2 - Developmental Process 101
   Motor development 102
   Passive carrying 105
   Partially active 106
   Active carrying 106
   Newborns to 12 weeks 107
   Shoulder hug 109
   Chest-to-chest 111
   Once upper torso control is achieved 112
   Transitioning to hip 113
   Independent sitting 116
   Crawling 119
   Cruising and walking 121
   Further development 122

Chapter 2.3 - Communication 125
   Physical communication in carrying 127
   Dyad and environment interaction 130
   Communicating needs 133

PART III 135

Chapter 3.1 - Physiology and Anatomy of the Baby 137
   Central nervous system (CNS) and neurodevelopment. 137
   Cardiovascular and respiratory systems 139

Head size 140
Upper body 141
Spine 142
Hips 147
Body fat 154
How babies' muscles develop 155
Legs 159
Feet 160

Chapter 3.2 - Differences Between Caregivers' Bodies . 163
Upper body 163
Hips 165
Muscle 167
Body fat 169
Postpartum bodies, lactating vs. non-lactating and pregnancy 170
Multiple caregivers 172

Chapter 3.3 - Side Preferences 175
Handedness 175
Caregiver preferences 176
Babies' handedness and side preferences 178
An impromptu survey 179
Active carrying and side preference 183

PART IV 187

Chapter 4.1 - Evolution, Primates and Carrying 189
  The "loss" of human hair 192
  A new concept 194

Chapter 4.2 - Frictional Properties of Skin and Hair 197
  Skin 197
  Moisturisers and other lotions 200
  Sweat and sebum 202
  Vellus hair 205
  Cold weather and goosebumps 208

Chapter 4.3 - Senses in Carrying 211
  Perceptual systems 211
  Sight 213
  Smell 217
  Sound 218
  Touch 219

Chapter 4.4 - Barriers to Sensory Input 227
  One layer on caregiver 228
  One layer on child 229
  One layer on each 230
  Underwear 230
  Socks and shoes 232
  Sight 233
  Sound 234
  Equilibrioception 234

PART V 235

Chapter 5.1 - Further Thoughts 237
   Weight in relation to physical development........................ 244
   Reaching developmental milestones................................ 246
   Caregivers' bodies.. 247
   Active carrying as a pleasurable experience 248
   Energetic Costs of Carrying... 249

Chapter 5.2 - What We Need from the Scientific and Wider Community 251

References 257

Acknowledgements

Many individuals have helped in various ways to ensure I completed this book and having people who are rooting for you makes a huge difference when you're walking into the unknown.

As always, thank you to my husband, Tom, for his ongoing support of my work. It's not always easy for him, especially during the process of me writing books and working away from home, but he continues to encourage me and celebrate my achievements. I'm forever thankful to my children for being a never-ending source of love and inspiration in my life. It's especially wonderful being able to share my work with my eldest, Niamh, and have her working and learning beside me.

A big thank you to Ulrika for being as obsessed as me about in-arms, keeping me motivated and passionate about sharing and learning more about this incredible subject. I treasure our discussions; some of which we record to share with the world via our podcast, Back to Basics: babies, bodies and behaviour. Her sharing of some of her thoughts on the possible evolution of carrying as well as knowledge of the early weeks of non-human primates' lives have proved useful in research for this book. I value her work greatly and look forward to her contributions to the collective in-arms knowledge.

A special thank you goes to Henrik, for his ongoing belief in me and my work as well as connecting me with people who may have interest in it. I am grateful for the continued support and encouragement I receive. My humble thanks also for writing the foreword to this book.

I'm grateful for the never-ending support of my wonderful friend, Mirrie, who is a champion of being kind to, and having belief in, yourself. Her positivity is contagious, and she's a pure joy to be around. She is an inspiration, and her passion, motivation and accountability – along with sticking around for both the highs and lows of the writing process – have played a huge role in this book coming to fruition.

My gratitude also goes to Nicola whose support and guidance was a big catalyst in ensuring I finished this project. The one to see the bigger picture when I'm too close to it, I've gained so much by being around her positive influence.

Input from people who have specialist knowledge and those who come across bits of information which may help shape our understanding are always greatly appreciated. I have received valuable information from Aiki, Sara and Shaughn which have helped shape parts of this text as well as my understanding of various topics related to in-arms, for which I'm most grateful.

Ongoing virtual support by way of interactions and appreciation of my work online never goes unnoticed. There are many people from which this comes from, but in particular, Zoe and Kimmy are two people I would like to acknowledge. It's the sort that keeps you feeling heard and appreciated in a beautifully familiar way, even if our work has different focuses. There's also something magical about women supporting women, and these two continue to inspire me with their work also.

I'm thankful to all the people who have recognised clinging as a normal human behaviour since the release of my first book in 2017 – the acceptance of my work and interest in the concept has contributed greatly to me sharing this latest offering, and I hope it is as well-received as the first. I'm also grateful for the lessons I learn from challenging experiences and interactions in my mission to make the knowledge of human babies as clinging young widespread and easily available. Although we tend to hold onto and place more value in the ways in which we're uplifted, I can honestly say that part of the reason I've achieved so much is because of these learning experiences, which always seem to come at just the right time. They – just like positive experiences – provide fuel for growth and creativity, which is a wonderful thing!

Last, but certainly not least, I'm eternally grateful for the wonderful people who conduct research into and form theories about human behaviour, development and evolution. Every bit of the research I have sifted through has helped to shape my findings, whether or not it was included in this book. Research helps with backing up our anecdotal findings and thought processes, enhances our understanding of relevant areas, and will sometimes give us reason to question and form our own ideas and theories going forwards. In an age where access to information is increasingly easy, I'm thankful for the ways in which independent researchers are able to get hold of the materials they need to learn more.

Foreword

This is a very important, and dare I say, revolutionary book for the future development of parent-infant contact practices, and I urge you to read Mel Cyrille's brave, compelling and riveting exploration of hitherto uncharted land.

When I first heard Mel describe the human infant as an active clinger, contrasting it with our predominant view of the infant as someone who passively needs to receive our carrying, my thoughts strayed off to a historic parallel in the world of medicine and psychology.

All the way up to the 1970's, infants were considered to be insensitive to physical pain, let alone emotional pain. This, we now know, fundamentally ignorant and wrong view of the infant's competencies and capabilities had devastating effects on how children were treated medically and socially for centuries. However, through the ingenious, respectful and most detailed research by especially two Harvard University researchers, Barry T. Brazelton and Daniel Stern, the scientific case was made that from birth onwards the newborn infant is a social, intentional and contingently communicating being. In addition, the infant is not just, as hitherto thought, a passive recipient of social stimuli from its caregivers. On the contrary, the one to two months old infant is an active initiator of real, ravishing, even hilarious proto-conversations, which brings so much reward and meaningfulness of parenthood to its caregivers.

The international grassroots babywearing community has made an incredible impact on modern caregiving over the past two decades, highlighting amongst perceptive parents the importance of extended physical contact and the negative effects of too pervasive physical separation. Parallel to this development, research in just the very recent years is ever more clearly demonstrating the long positive reach of early extended physical contact, with both cognitive and socio-emotional effects in late childhood. Slowly, we are learning the truth of preeminent paediatrician, Donald Winnicott's statement from the 1930's that "There is no such thing as a baby. There is only a baby and its mother". These days, we would add "or its father, father figure, other family member, caregiver". Nevertheless, we begin to understand that a separated baby is a completely different baby to a carried baby, whether viewed physiologically, behaviourally or psychologically.

For all this important progress, we in the babywearing community just might have had a similar blind angle to the one described for medicine and psychology. In our attempts to respond to historic widespread medical concerns about the effects of babywearing on infant spine and hip development, with the consequent focus on correct, ergonomic positioning in a variety of carrying devices, we may have missed a deeper truth about the infant's inherent capability to actively contribute to the carrying process. By ignoring this, we may inadvertently bar the infant from important early behavioural experiences, with potential deleterious effects on their agency and embodied sense of competence in dealing with both other beings as well as the physical universe.

Ironically, Dr. Evelin Kirkilionis did in fact in the 1990's demonstrate in various ways, how an infant is evolutionarily programmed to be actively carried, through its own adaptations and behavioural contributions to the carrying process. But her work was not taken to its natural consequence, but instead formed the foundation for correct positioning in carriers.

Similarly, groundbreaking observations by Swedish physiology and midwifery researchers in the 1990's made it crystal clear that even the newborn does not in any way need help to be put to the breast. No, when placed at the lower abdomen immediately after birth, the newborn will go through nine distinctive behavioural phases, which include active crawling up the mother's body to reach the breast. These observations have furthermore led to the formulation of "biological nurturing" where the infant is given opportunity to exert its own behavioural agency in the feeding process.

Drawing on these research impulses and experiences with both her own three children as well as numerous clients, Mel makes a strong case for a fundamental revision of our view of the infant's capacity to contribute to the carrying process. When viewed in this light, the actively capable clinging infant necessitates a new understanding of many anatomical aspects of infants, mothers, fathers and other caregivers, how we engage in the carrying process, the role of skin-to-skin contact and our own preconceptions about what comfort for us as infant caregivers really is.

Importantly, I do not in any way read Mel's book as an attack on current babywearing practices, but more as a paradigm-shifting perception of the infant, which will necessitate adjustments in how and when we use carriers and ultimately also in carrier designs.

As for any fledgling research strain, there are still many questions to be answered, and Mel is refreshingly honest about this. But with her book she has certainly won me over that any infant should be allowed to develop their own specific capability for active clinging and brought about a curiosity as to how one may best support that, with many useful advices provided here.

Henrik Norholt, PhD

Chief Science Officer
Ergobaby Inc.

Introduction

When I wrote my book "In-arms Carrying: A practical guide to comfortable carrying" in 2017, I had multiple occasions of frustration which went beyond my disorganised personality traits combining with the unreliability of technology. Many of my frustrations came from needing to write a book that was "basic" enough to introduce the concept to both caregivers and anyone whose interest was piqued by the idea of carrying being more than holding. As a new concept, in the way that I perceived in-arms carrying versus how it was known in the world, it needed to cover all the introductory topics and be as easily accessible as possible. This left me a bit deflated about having so much more to share, but nobody to share it with.

Thankfully, the amazing reception to my ideas in the almost-12 months since the release of my first book has encouraged me to write a second one (after swearing never again). Something I thrive on is gaining and sharing new information, discussing with others and forming new theories and ideas in the process. This is one of the main purposes of this book – to instigate more talk about the scientific and theoretical sides of carrying and encourage people with specialist knowledge to take these ideas further and bring about a new normal in child development. Also, so I may selfishly learn more! As I always acknowledge, I am but one person, and a person without deep specialist knowledge in certain areas. There comes a point where it becomes difficult to develop your theories further without the relevant insights that specialisation offers.

This book is focused on many of the (seemingly endless) ways in which our bodies and those of our babies and children are inherently designed to make in-arms carrying both possible and easier to do. Babies are born to be clingers, and caregivers are made perfectly to support the carrying process. From the bowed legs of babies and young children, to the fascinating properties of skin and hair, to the way our bodies are structured - if we look at the human body from a carrying viewpoint, so many revelations are made, and I'm sure that many more will follow in the years to come!

This book is mainly aimed at the avid enthusiasts of active carrying, specialists and educators. To keep this book at a readable length I've covered what I believe would be some of the most interesting aspects of clinging behaviours and compatibility between bodies. I wouldn't say this is a complete guide to the science behind clinging as there is much more I haven't covered and no doubt unfathomable amounts yet to be discovered.

I hope you enjoy this book, whatever your experience with carrying and whatever your expertise in the areas related to it. If you want to connect, please do so by emailing me at: hello@melcyrille.com – I'm always open to chatting with enthusiasts – my passion for in-arms carrying is limitless!

Mel

July 2018

PART I

Chapter 1.1

The Clinging Young Concept

It's a widely accepted fact that human babies need to be carried. In the simplest of terms this is because – in comparison to the older child and adult – they are born significantly underdeveloped and rely on a caregiver to complete tasks for them to ensure their survival. Caring for infants requires moving them from one place to another, feeding and comforting them. These things are significantly easier to do in-arms. Knowing the scientific classification of human babies provides us with valuable insight into both their needs and how they are designed to behave on-body.

For a very long time human babies were thought to be "secondary nest dwellers" of the altricial young category of animals. That is, immature, helpless nest dwellers who rely on a caregiver to bring them food, but who cannot be left for long periods of time. In 1970, Bernhard Hassenstein described a third category of mammals to differentiate humans (as well as apes, bats and marsupials) from the altricial nest-dweller and precocial parent following young.[1] He named this category "Tragling", meaning "parent clinger". The description of this type of young was that they are preadapted to hold on to the caregiver in the infant stage of life, with the primary caregiver's body being the natural habitat for them. In 1992, Evelin Kirkilionis studied human carrying behaviours further and described human babies and children as "active clinging young".[2] She had found that we are still born with the ability to physically cling to our caregiver, therefore sought to bring forth an understanding of this.

Whilst the findings from various studies seem to make sense, in modern society we don't tend to see babies as active clingers. In fact, in many places where human carrying is discussed, it is regularly stated that babies "lost their ability to cling" to their caregiver.[3] It seems that this ideology fuels some confusion surrounding Kirkilionis' concept. It seems to get lost in translation and is widely used to promote the idea that human babies are adapted to be worn in baby slings and carriers. Whilst this is also true, the main focus must come back to the original definition – babies are designed to *cling* to the caregiver.

The idea that the mother's body is the infant's "natural habitat" makes complete sense; especially if this is the gestational parent and primary caregiver. This is the body which was designed to be the food source, in the womb via the placenta and outside, through the breasts. The person who grew the baby is biologically expected to form the first attachment to them soon after birth, and it's on their body that this is designed to happen. Post-birth hormones, touch, taste, sight, sound and so forth work together to ensure the baby recognises this as an extension of their first home and that this person develops an emotional attachment to them, ensuring they will be looked after. The physical side to all of this is directly related to the fact they are designed to be carried for a large part of their life in the first year.

We're going to explore how human bodies are adapted to both cling to others and to support active carrying. We'll also discover biological differences of male and female bodies which impact on carrying. Human clinging young is evolving. We live in a time where families of a one male, one female set up are becoming less and less, and more people are being born into bodies they do not identify with. For the species to adapt to these changes we must have an understanding of the biological baseline for carrying. This book seeks to bring life to the concept by bringing together many different aspects of science and biology which confirm the undeniable fact that human infants are active clinging young.

Chapter 1.2

Introduction to How Clinging Works

We know that clinging behaviours are formed in the womb, long before they present in clinging as we understand it. This is by way of the primitive reflexes, and the ones involved in newborn clinging are used for many different actions and other forms of movement. For example, the palmar reflex is seen when babies grasp their umbilical cord or hands in utero. Others are responsible for positioning (e.g. Tonic Labyrinthine Reflex in flexion) in a way which makes the most of the confined environment of our first "home". Some assist in the normal birthing process (e.g. TLR in extension). Reflexes are designed to be stimulated for different reasons to create movement patterns to assist the infant in navigating their environment before their motor development has reached a point where voluntary actions are possible. It's no wonder that missing the biologically essential movement processes can hinder physical development going forwards.[1] It raises important questions about whether other normal developmental processes can fill in the gaps left by deviating from what is biologically expected of us, or whether completing every process in specific ways is required.

Clinging relies on the baby using its inborn reflexes to aid the carrying process in the early weeks and months and the integration of these at differing times enables them to learn voluntary clinging bit by bit. In this chapter we will explore certain aspects which aid clinging and will go into further detail as we progress through the book.

Critical period hypothesis

We know that there is a biological baseline for normal human development, whether that is facilitated by normal infant nourishment, movement, physical contact, and more. There are theories that there are critical periods for both normal and optimal physical development.[2] These periods are times where specific experiences have a greater effect on development than other times, rather than being absolutely essential for the emergence of a skill.[3] Although there is evidence to back up these theories, there is little understanding of the specific timeframe of such windows of opportunity.

The critical periods appear to start soon after exposure to the relevant stimuli. This is combined with a readiness for the stimulation to be effective and the behaviour to be developed – they must have already acquired a certain number of attributes which will make the next step possible. An interesting part of this theory is the idea that the effects of stimulation during such periods create lasting effects on the infant.[4] This means that even if the stimulation wasn't enough to achieve optimal development, the fact that *some* was given means there is an opportunity to revisit the skill at a later date and develop a mastery of it to sub-optimal genetic potential. The ability to play "catch up" after critical periods pass appear to be an inbuilt "safety" feature which enables humans to get back to a relative normal. Plasticity of the brain continues, though it may not be at the same level as during the critical period. On the one hand, it is frustrating as it appears there is for each skill a period of time where optimal achievement can be reached based on the person's inbuilt spectrum of potential. On the other, it's very useful that a person can revisit said skill at another time in life and master it, albeit possibly at a lesser capacity. In carrying, this means babies are meant to develop clinging behaviours at a certain time in life. If some of these behaviours are stimulated but on-body clinging is not encouraged, it will still be possible to relearn the skill after the reflexive periods have passed.

As you can imagine, clinging – along with other developmental processes – goes through many critical periods. Readiness to develop specific clinging behaviours is not present at the same time for each of them, and this is a wonderful thing as each behaviour gets the attention it needs. This can mean a child may show different levels of abilities of various actions associated with clinging. For example, a baby who has been carried in more passive ways is likely to show good upper body stabilising abilities if it was just the lower body which had been restricted. Similarly, when off-body clinging behaviours are developed normally, they can be applied on-body. Many of the movements in active carrying are practiced off-body, especially in the reflexive periods. However, learning the corresponding behaviours during the critical periods ensures smoother transitions and full use of the actions from the point of readiness, and helps the baby to learn how to apply them to the on-body environment.

How a baby clings

As we'll explore in greater detail in the next chapter, newborn clinging works by activating many different – but inter-woven – primitive reflexes. These reflexes make up the foundation for active and voluntary clinging to be learned as the reflexes integrate. Clinging reflexes are elicited by various sensory triggers, such as tactile and vestibular stimulation. As we'll discover, they work at their best in groups. Each elicits a reaction in their own right, but when they join together they produce a larger movement or more complex action. Clinging requires multiple contact points and specific behaviours to produce the end result. This all changes based on whether the caregiver is providing any support, and where that support is being given. It also differs based on the baby/child's current clinging abilities. Clinging behaviour in-arms also looks different based on the amount of bodily contact the child has. Full-body contact tends to signal passivity (except in the case of independent clinging), and partial contact indicates activeness. For the legs though, full contact is required by them to perform the clinging action. The more leg in contact with the caregiver's body, the greater clinging potential.

Babies and children are able to cling with the feet also, and thigh, calf and foot contact elicits the strongest general clinging behaviour.

Some newborn clinging behaviours work in a different way to older babies' ones, and this is due to the period of time when their muscles aren't developed sufficiently to facilitate active clinging. Instead, the reflexes and posture the baby adopts helps make them more portable and fit to the chest and off-centre by the shoulder (shoulder hug) better. In the early weeks and months, there tends to be a big focus on the squat shape a young baby reflexively goes into. This posture, combined with the clinging reflexes of the thighs and legs, makes clinging possible as they reach a point in their physical development where it's appropriate to transition to the hip. The fact it is also elicited when laying down suggests that contact/pressure on the back may also be a possible trigger. Clinging also occurs with the grasping reflexes of the hands and feet on the carrying person's skin and/or clothing. Babies practice these carrying movements as their reflexes are triggered in more supportive positions as their neck and upper body muscles develop. Once they're strong enough, these movements can be applied to active clinging positions. The clinging reflexes truly come into their own once active carrying begins on the hip.

The move to the hip position is where we begin to see clinging in the true sense. The legs and feet work in a different way to the earlier shoulder hug and chest-to-chest positions, and the baby finally has a surface on which it can slot onto and grasp with their legs. The baby has a new – but familiar – environment to navigate, and the presence of carrying reflexes help make this transition easier. Reflexive clinging gradually develops – bit by bit – into voluntary as awareness and control take over. As it evolves we will see changes from different parts of their body over the months and years as their body grows and their weight increases. As we know from observation (and in combination with the critical period hypothesis) a baby or child must have the opportunities to participate in active carrying at a development-appropriate level during certain periods of their life. If they don't the ability is more often than not lost or greatly weakened.

Position of legs

What position their legs are in affects the child's ability to cling. They're able to work with the caregiver's body in different ways in different positions, so it's not as simple as saying, for example, that being in a slight squat position helps in all holds. Leg position needed to facilitate clinging will also vary depending on how far they're into their physical development, how long their legs are, and how the caregiver's body shape and makeup impacts on it. Also, when a baby is new to active clinging with their legs they are using the different parts of their extremities in different ways to an older baby who can cling in a voluntary manner. This isn't to say that when they're younger their clinging capacity is worse, it's just that it matures in relation to physical development, which is needed as they grow bigger and heavier. If clinging stayed the same throughout the carrying years it would become harder for the caregiver to support them and harder for the baby to cling. Remember that physical development starts from the head and works its way down to the feet. As voluntary behaviours are learned by the legs and feet, clinging can be aided in new ways.

Younger babies tend to cling in a deeper squat position in a hip carry.[5] A reduction in angle appears to be linked to the integration of the squat reflex. The need for a deeper squat may also be linked to the baby's hip position in relation to their spine at this time as the hips move backwards as the lumbar curve develops. Also, the higher the legs, the more curved the spine, and in a more developed active hip carry there is marked detachment of the dyad's torso. As a baby achieves independent sitting it's much easier for them to maintain a more seated, detached position. When voluntary clinging with the legs has been achieved, full body contact weakens their clinging ability as it relaxes the body. A detachment requires them to engage their core muscles and hold their torso upright.

When children's legs are long enough they're able to use their ankles and feet in different ways to aid clinging. They may, for example, use their heels to hook around the caregiver's sides to make an anchoring point in front and back carrying. This means their thighs don't need to do as much work. As the clinging surface is wider on the front and back of the body in relation to the side/hip, the baby must work harder to cling on. This tends to be easier on the front of the body due to the many variations in support possible from the caregiver's arms. On the back, it's much harder for a baby to cling until they reach a height which enables the use of their arms over the caregiver's shoulders. This is because the orientation of the carrying person's arms is restricted and support of the baby's back is near impossible. This means back clinging tends to emerge later than front clinging.

Big children are also able to cross their ankles in a hip carry, again locking themselves onto the body in a way which eases the active work needed by the rest of their legs. In front and back carrying they can use their calves to hook around the opposite side of the body, knees bent at the caregiver's sides. This may indicate that there are certain points in relation to weight and development where the maximum capacity is reached in terms of the amount of time being carried each day/week/month in relation to their increasing weight, and clinging must evolve to support longer clinging periods. Without position-specific practice the clinging muscles won't be kept at their optimum level. This different way of using their body to assist clinging is a handy adaptation to facilitate the clinging of the older child.

Something which impacts leg position is how much skin is exposed on them, as well as if carrying is taking place against a clothed caregiver or their skin. In fully-clothed carrying we tend to see a greater disconnect of the lower legs and this means clinging must happen with just the thighs. This obviously requires more work from one area of the body, so they tire more easily. It can make carrying in colder weather harder to do when out and about, as well as the fact there's likely to be more bulk, which is also an interference. However, it's not as though learning clinging behaviours is completely disrupted. When babies and children are younger they tend to need to be picked up a lot at home and carrying here in the warmth aids their clinging development over the colder months. When more skin is able to be exposed the baby tends to engage more of their leg in carrying, and this is likely due to the removal of a barrier to sensory input. We will explore the roles of skin and the senses in carrying towards the end of the book.

A workable environment

For clinging to work, the environment in which it occurs must be favourable for the clinger. We will only touch briefly on this here as it is covered in much detail during Part III. We are looking for things such as:

- Bodies which fit well together – pre-adaptations for carrying person and clinger (e.g. being able to sit on the body, caregiver's arms being designed to support various positions)

- Sex-related advantages (e.g. females being designed to facilitate carrying more than males)

- "Materials" which behave in a way in which facilitates clinging (e.g. skin, certain clothing fibres)

- Effective sensory input to make the process smoother

The human body is designed in very specific ways to make in-arms carrying possible. Working with these inbuilt mechanisms makes clinging easier to do.

Endurance

Something which naturally comes up as question is how is it possible for humans to cling and support clinging for extended periods of time without fatigue? For a start we must remember that – for the caregiver – a clinging baby is a much lighter load than a passive one. There is much less work involved in carrying them as the carrying person is merely providing a gentle support with their arm, the baby is doing the work to cling, and the rest of the perceived weight is being born by large muscle groups of the part of the body they're clinging to. The arm is doing significantly less work than in passive holds, as it's bearing minimal weight. Therefore, it's much easier for the caregiver to carry for longer periods of time. They also have the advantage of being able to switch up positions as and when is needed, allowing different parts of their body to support the child. For the baby, in the first half-year or so they possess an ability to hold the spread-squat position without fatigue due to the reflexive nature of the pose. This position is tiring for adults but not babies. As they begin to take voluntary control over this action they build up muscular strength and endurance to hold this position and similar (up to 90° angle). If they do not continue to use this position to cling they will find it harder to do if required at a later date. Their legs are also in constant use at this point, learning to navigate their environment off-body.

It's one of those things that can be hard for the caregiver to get their head around, as passive carrying can be so cumbersome and the idea of a baby doing much of the work may be hard to believe. Also, there isn't much we do as average human beings where we rely on ours and their endurance in such an engaged manner. When seen in action, clinging (and supporting it) is clearly very obviously doable and can be done extremely comfortably. Whilst these explanations make it clearer as to the role of the muscles in clinging and supporting it, another question remains: how do the muscles cope, both in caregiver and child?

Muscular endurance relies on muscles to contract repeatedly over long periods of time. Endurance is built up by doing an activity for longer periods of time, increasing the weight, or a combination of both. When we carry a baby from birth regularly, we adapt gradually to their increasing weight, therefore build endurance this way. However, it can become increasingly difficult to carry a passive load, and the lighter something is, the longer we are able to carry it for. When we work with the child's clinging abilities we enable them to decrease their perceived weight, meaning we are less likely to reach a point where complete fatigue happens, and we cannot support active carrying anymore. We also build overall endurance by switching sides we're carrying on, which substantially increases the possible carrying period.

For the baby, when clinging is used in carrying it's also not produced at maximum output. We know this because of their ability to cling harder if they slip or don't want you to put them down. This means a lower rate of fatigue and the differences can be seen in action when less than needed support is given. The baby clings harder, trying to hold on well enough, and fatigues easily. This suggests that maximum output relies on different muscle types (as does passive carrying) and is another area we will explore further in Part III. This is similar to how we can carry a light bag for long periods of time without it bothering us but carrying a heavy one has a definite point of fatigue.

Carrying is more of a muscular endurance task for the caregiver when a baby is clinging than cardiovascular, but this is also a factor in endurance activities for the child. Babies are not designed to endure cardiovascular stress, and when we observe clinging in action it's clear that this is not a stressful activity for them. They can easily hold a conversation and maintain a regular state of being, similar to when they're held passively. Cardiovascular fatigue is reached significantly faster in passive holds for the carrying person, especially with a heavier baby/child. To carry for extended periods, such as going for a long walk, the caregiver must not be worn out by the additional load. How easily they fatigue will depend on other factors too, such as if they are carrying additional bodyweight and what their normal walking capacity is, but if they are using carrying as a normal daily task endurance will be built naturally, just as for muscles. Again, this is why it is important for clinging to be developed – a clinger feels lighter than a passive load, so cardiovascular demand is less in active carrying.

Variables in carrying environment

Not all carrying will be on a flat, level surface, although many of us spend most of our time in this sort of environment. Even in cities and other places with a lot of pavements we're likely to encounter some uphill and downhill situations, along with steps and stairs. The natural world is full of many differences in ground texture, inclines, declines and so forth, and is the normal environment for carrying to occur in. Of course, for many of us the man-made environments have replaced this, and so natural terrain is encountered less and less. Let's now take a look at how clinging can be affected by the caregiver walking on different surfaces and at various angles.

When we're walking uphill, our body changes position so that we're angled in a different way. The torso leans forward slightly, and the thighs appear more angled, creating a potential on-off bracing point. This changes how a baby must cling as a forwards-leaning body in and of itself is counterproductive in carrying. However, the change in stance triggers a difference in clinging behaviour and the child is able to adapt fairly easily when clinging. In a hip carry a forward lean means the dominant leg (to the front of the caregiver's body) is able to cling less, but the non-dominant one is triggered to brace to a certain extent. The forward motion also tends to trigger a stronger grasping and clinging of the upper body and hand/s. For a baby with long enough legs, they're able to use our legs to their (and our) advantage. They're able to connect with our leg closest to their foot on the front of our body and push on it as it comes forward. At first thought this might seem counter-productive, but it isn't. The contact comes as the leg is in flexion and the push occurs as the leg travels through to extension. Also, the push tends to be fairly subtle so it's not as if they're bearing their full weight on you, making a harder job even more difficult. The leg is able to briefly bear a portion of their weight at short intervals, which aids the carrying process. In back clinging the slight forward lean is particularly helpful for the toddler who is developing their skills in this new area.

Going upstairs is a variation on the uphill. Significantly, the leg goes from flexion to extension on the upright rather than the diagonal. This makes it harder to do in hip carrying, but it can still be fairly straightforward in back clinging. For hip carrying we may see a drawing of the baby to the caregiver's body in compensation for the change in gait. So, how about walking downhill? Naturally, we walk with shorter and faster steps as we attempt to counter the acceleration that comes with this change in angle. In carrying, it's common to see a slight backwards lean of the body, which of course provides the clinger with a little relief from the faster movement. This may seem counterproductive to the caregiver, but the change in speed along with stride also produces a stronger clinging response from the baby. The slightly reclined body provides a buffer from the acceleration which helps stabilise the child.

Different surfaces also affect carrying in different ways and we're going to look at two types of ground. Stones can produce different reactions from the caregiver's body. If we're walking barefoot and our feet aren't fully used to walking on stones, we may sometimes react with a withdrawing action. This reflex jerks the knee up to a certain extent (depending on the amount of pain felt) to remove the foot from the sharp object. This is helpful in carrying as sudden movements tend to elicit a tighter clinging response from the child. The jerking reaction of the arms too helps with drawing the baby closer to us. In a general sense though, walking on stones or other bumpy ground sends various "mechanical nutrients"[6] through to the baby. They experience a more well-rounded carrying experience from all of the micro- and macro-variations in the caregivers walking environment.

Another interesting natural surface is sand. Walking in sand is harder work than walking on a pavement. The moving surface as we step (moving the grains of sand) means we our feet and bodies are moved in different ways. It's interesting that beaches are often associated with swimming and sunbathing, which tend to require less clothing. In warm weather they can be a useful way of incorporating some skin-to-skin clinging whilst walking on a more difficult surface.

Carrying is movement

It's widely accepted that one of the reasons babies are designed to be carried is because of an immaturity resulting in the inability to escape predators. The idea which goes along with this (and has been held up by some research[7]) is that babies need to be calmer and stiller to make it easier for the caregiver to carry them. Whilst this may very well be true, there's another angle to look at this from. Babies *expect* to be carried. They've been carried in the womb up until birth and the stillness of the new world around them is a stark contrast. It wouldn't make much sense if they immediately adapted to not being carried. Carrying is movement; by way of the carrying person *and* the baby/child. Babies expect and need this movement. This is why they require a combination of physical touch and movement to achieve an optimal environment for them to thrive in. This goes beyond carrying, of course, but this touch and movement is our focus here. It's not always enough to just have the touch of the person who carried them in their womb, nor is it enough at times to be carried (receive movement) from another caregiver. It's no wonder then that being carried by the primary caregiver tends to be the environment many babies are happiest in. The person whose positive touch they are most familiar with can bring about an optimal environment for carrying.

Movement from the caregiver also produces a trigger for clinging and it tends to be noticeable that babies participate less when we stand still. If babies need movement and this promotes clinging behaviour, it seems natural that the next revelation would be that they also expect and need the chance to participate in carrying too. Whilst this has yet to be scientifically proven, it is clear when observing active-alert babies who are used to being carried actively by their caregiver that they tend to signal (verbally or non-verbally) their desire to participate. It's interesting to observe younger babies who have yet to be fully conditioned into expecting passivity in carrying. We tend to see more bodily movement when in the active-alert state, almost as though they are physically communicating their need for participation in carrying.

Clinging requires the participant to be active and alert. A tired baby does not engage as well in active carrying and is likely to get frustrated or upset if the caregiver is trying to instigate it. So, timing is key. Passive holds are most useful for times when they're tired or sleeping. This ties in well with active carrying being movement for the baby – when they're fatigued they have little desire to engage in exercise. Passive carrying is also movement, but in a different way, and the movement may come in different forms based on how the baby is being held. For example, a child's legs being supported by their caregiver's arm are passive as they're sinking into the "seat" provided and aren't working to cling, yet the movements of the carrying person gently move the baby with each step. One effect of this is varying pressure on the joints of the hips, controlled by the force generated from the impact of the caregiver's arm.

The movements of the baby's body range from imperceptible to obvious. The smaller movements happen in the joints, muscles and skeleton and help with spinal development, building strength and the loadbearing contributes to the ossification of joints.[8] Carrying provides them with mechanical nutrients in a different way to when they interact with the world off-body. The larger movements that we see tend to be for positional correction, communication and orienting themselves to engage or disengage with the world around them. Many of the movements involved in clinging are happening where we don't see it. For example, countering the carrying person's movements (e.g. walking and turning) requires the baby to engage their core muscles, but this is not obvious to the naked eye. We can, however, feel this happening if we're to place our hand on their abdomen. The movement that is carrying is fascinating and its difference to other movements (such as crawling and walking) is something I believe will reveal many interesting insights once researched further.

A passive society

In our society's quest for "advancement" we've left ourselves way behind physically by inventing ways of making life "easier" and focusing on outsourcing movement and connection to make room for the demands of modern life. We're living in a time where we absolutely need to recognise the impact of our changing needs to function at a level which enables us to keep earning an income needed to survive this new climate, but also takes into consideration the negative impacts of such a society.

One such negative impact is, of course, the way we have come to carry our babies and children. Even though babies are preadapted to cling to their caregiver and display these capabilities quite clearly, it can be hard for caregivers to recognise them. In England, the impact of Queen Victoria using perambulators[9] brought about perceived "advancement" as the general public bought into this method of transportation, but this has inadvertently led to the detriment of infants due to a growing trend of "containerisation" of babies.[10] Human infants in England tend to be carried much less than they are designed to be. Although there has been a strong current of change in parenting advice in recent times – arguably with thanks to a plethora of research showing the importance of responsiveness and touch from caregivers – there is still a great reliance on baby containers. On the one hand there are the clever marketing techniques drawing parents in, with the belief that products will help their child to develop at an enhanced rate. On the other there's a greater reliance on ways to keep babies both occupied and safe whilst caregivers get on with all the other things which need doing in their busy lives. There are – of course – others who depend on such devices as an aid due to health restrictions, but I am addressing here the majority population.

In many predominantly pram-filled cultures babywearing is making a comeback, which is helping parents and carers turn the focus back to holding babies close rather than keeping them at a distance. This is wonderful and is having a great impact on countless families. Although slings and carriers are still containers, they have many additional benefits due to the bodily contact that comes with using them. Now, though, is the time to take this a step further and rediscover how important it is to let them also flourish as the clinging young that they are. We were *born* clinging young. Yet we are raised – from both a conscious and subconscious place – to not fulfil our inborn potential. Societal influences impact caregivers on a subconscious level. For example, here in England it's common to see the carrying person stick their hip out and hold up the front leg of the baby on their hip, even when the baby can cling to their hip with no postural adjustments needed. This is just one example of how babies are unintentionally set up for a sedentary life.

Most of us in Western society need to move more – there's no denying that. We have outsourced so many daily tasks that we truly are sedentary. It's no wonder that babies and children are seen as "whirlwinds" and adults find it difficult to comprehend where their seemingly boundless energy comes from. We've been trained, from an early age, to be still, move less, and "fit in" as mini adults. Looking at baby/child behaviour and movement as an example of normalcy throws up the question of "why do we not move in the ways children do?". Of course, the lack of "boundaries" and the impulsive nature of children tend to dictate to us that everything else which goes along with it needs "training" and "refinement" to make them functioning members of society. Following their movements and learning from their very nature (which is inbuilt) can aid our understanding of what health and wellbeing should look like as they grow into adults.

What does all this have to do with clinging? Well, encouraging passivity could well have a knock-on effect of clinging having less appeal to the baby, but more importantly (and more easily verifiable) is the impact the caregiver's carrying habits have on the child. Simply put, if active clinging opportunities are not given, the baby does not have the chance to develop on-body clinging behaviours. There can be things which may help them to partially develop the skills (such as babywearing and climbing), but passive carrying eliminates the normal and natural occurrence of the on-body clinging developmental process.

Next, the way in which a baby clings is determined by many things and one of those is how the caregiver holds their own body. In a world where technology is taking over, and the use of it starting in infancy, our posture is being affected more than ever. Along with the containerisation of our young, the majority of children are required to sit at a desk for large parts of the day at school. Add to this the increasing use of mobile phones and computers in work and leisure, we have a toxic mix of harmful sedentary behaviours happening each and every day. We're seeing an increasing trend of hunched shoulders and forward head posture, starting in childhood.[11] Sitting on chairs and sofas is encouraging pelvises to tilt backwards. A caregiver's hunched shoulders change the baby's position in clinging because of how a less open chest interferes with the available space on-body and impacts on how and where they're supported. A tucked pelvis provides extra support for the baby on the front of your body so encourages less clinging, which in turn makes the carrying person's body have to bear greater weight. Using the body as a bracing point makes for a greater risk of injury.

Whilst misalignment can't always be corrected, and we will have to find ways of working around this, knowing how clinging is impacted helps us to understand the habits babies and children develop that are linked to them (e.g. seeking passivity in carrying), and sheds light on the deviations from biologically normal clinging behaviours. This is yet another reason why it's so important for us to be trying as best we can to raise a generation of children with bodies closer to normal than the last. Not being able to move our body in normal ways affects our health and impacts on that of future generations.[12] Babies are usually born with bodies designed to move normally. It's later on that issues arise, but this can happen earlier on in life than we may expect. Babies need regular movement and the movement which is carrying is as natural as rolling or crawling.

Clinging behaviours disappear due to passivity in carrying, so for clinging to work the caregiver/s must give the baby opportunities to be an active participant. Voluntary on-body clinging rarely appears without specific work to develop it. Active carrying is a movement easily added into a baby's life, as carrying happens regardless. All the caregiver need do is ensure they change the way in which they support the child. The thing which may require more effort is to switch up the places in which carrying occurs, as caregivers tend to have a side preference, which we will explore in detail later in the book. Moving more and in varied environments is important for both adults and children. As I touched on earlier, walking on different surfaces (ideally those found in nature) can enhance the baby's carrying experience. They learn to adapt their clinging from the effects of walking on uneven surfaces and changes in gradient. It gives them a well-rounded experience and the carrying person receives multiple benefits too.[13]

Chapter 1.3

The Role of Reflexes

Babies are born with an amazing array of reflexes – some of which will stay for life (such as breathing and blinking) and others which will integrate over time and be replaced by voluntary actions. Many of these are also carrying reflexes. The primitive reflexes develop in the womb and their initial purposes include foetal positioning and consumption of amniotic fluid. They're also essential for the journey from the womb, through the birth canal, to the outside world, reflexively navigating their way as certain responses are elicited. Reflexes are intended to be activated at certain times for certain purposes and disruption of these processes (including missing out on the normal birth process) can lead to increased risk of developmental delay.[1]

Reflexes are designed to produce reflexive movement long before the baby has the muscular and cerebral development needed for certain voluntary actions. Imagine being new to this world, outside of the womb. As someone with very limited vision, muscle strength and motor development, you seem quite helpless. However, you are incredibly intelligent, even if it may not be that noticeable to your caregivers. You've been hardwired to interact with your environment in a reflexive way and have many survival instincts. You are not helpless, and if your caregiver is aware of the various ways you are able to communicate, and what your inborn skills are, carrying becomes easier for you both.

As you can imagine, to create the movement which is "carrying", babies need to develop fluid motion and quick reaction times. This is where reflexes come into play in the beginning, and why some reflexes will still be able to be triggered in extreme circumstances once they have been replaced by voluntary actions. The repetitive movements gradually become smoother and better coordinated, which seems to be indicative of why active carrying happens in a different way in the first few months of life. The baby is designed to be in near-constant contact with their primary caregiver in the early weeks. The primary caregiver (usually the gestational parent at this time) is the infant's "home", and the carrying reflexes initially help them with getting to know this new environment better. As they grow and develop they're able to use their newfound strength and awareness alongside the primitive reflexes to implement their reflexive clinging in new ways.

Reflexes require a trigger to elicit a reaction, and this may be via movement, touch, disruption to balance, and other sensory stimulation. Many of the carrying reflexes are tactile ones - they rely on either touch, pressure or movement against the skin to trigger a reaction. A great deal of the lower body ones are superficial, relying on receptors in the skin to elicit a response. When reflexes are tested by medical professionals they're assessed separately and are usually conducted with the baby in supine or prone position. The focus is on the individual responses rather than how they can work together to produce various movements. They also tend to be tested in a way in which elicits an extreme reaction, which is understandable as a clear response is better for clinical evaluation than a subtle one. Missing or retained primitive reflexes are an indicator of neurological dysfunction, which is why assessment takes place in the newborn as standard. Older children and adults may be assessed when there are learning difficulties present or a diagnosis of being on the autistic spectrum. Very premature babies will be missing some reflexes as they develop at different times in utero.

In carrying we find that reflexes vary in strength depending on how they are stimulated and what other reflexes are doing to either aid a particular movement or inhibit part of a response. They also don't tend to be elicited for the main part in such a separate and non-fluid manner. Reflexes work in groups to produce the movements needed to make the baby an active participant in the environment they're in, and we see this very clearly in carrying. Whilst I've found little material discussing reflexes being linked together to create a chain reaction, it has been recognised in the reflexes of the organs.[2]

For the newborn, the reflexive movements exhibited are a continuation of those they made in the womb. The difference being that now they are out of a fluid-filled environment with very specific boundaries and are experiencing the full effects of gravity in a dry, spacious world. It will take some time for them to get used to this change in living space, but the muscular development which comes with the extra effects of gravity and freedom of movement contribute greatly to the development of control and integration of movements. The infant learns to counter the effects of gravity and in the process develops smoothness and coordination of bodily movement. The presence of the reflexes is reliant on both the age of the baby in question, and their current state of alertness. They tend to be elicited best in the active alert state, which is most conducive for active carrying, and some are absent during regular sleep.[3] Proprioceptive reflexes (those connected to position and movement of the body) have been found to elicit good responses during regular sleep and when a baby is awake, but are absent or markedly diminished in irregular sleep. Exteroceptive reflexes (elicited by external stimuli) are absent or diminished in regular sleep. During irregular sleep these reflexes can be elicited almost as well as during wakefulness. Nociceptive reflexes (responses to pain) are present in all three states.

Developing the nervous system

The central nervous system (CNS) is responsible for processing sensory stimulation and producing motor reactions in response. It's comprised of many parts, including the brain, spinal cord and nerves. As the CNS develops, more centres are developed which improves interpretation of the stimuli and allows for more refined actions. Primitive reflexes stimulate the nerve nets of the basal ganglia, whose main task is controlling motor activity, and this development leads to postural reflexes emerging. Movements improve in line with muscular strength and control, as many reflexes aid muscular development. For the CNS to develop it needs sensory stimulation and for motor development to improve there must be development of the CNS. Put simply, the reflexes must be stimulated for them to integrate, which in turn helps the CNS to mature appropriately. This is one reason why reflexes which fail to inhibit cause developmental delay. It can also be linked to why babies fail to thrive without touch – lack of tactile stimuli stunts physical development, just as a lack of positive touch can stunt emotional development.[4] Appropriate stimulation of the reflexes helps with the normal development of the CNS and enables voluntary actions to emerge. An intact nervous system is required for both the refinement of movement and the sequential development that takes place in a healthy baby.

For these reasons it's important that the baby receives adequate opportunities to move and interact with the world around them, in ways in which are developmentally appropriate for them. This is why we always look to developmental progress rather than age to determine such opportunities, and knowledge of the reflexes can also aid with ascertaining stage of development. As progress is made with physical development we can identify readiness for different positions and levels of support.

There are many primitive reflexes present at birth and a large number are involved in the carrying process. This is why being carried in-arms can be classed as an interaction with the environment (caregiver's body) and is an opportunity for movement. Here we will explore some of the carrying reflexes, and we're going to break them down into individual responses, then uncover how these work on the body and discover why they're so useful in carrying. Next, we'll bring them all together to understand how they're linked and why reflexive carrying is so intricate.

Plantar grasp and Babinski's sign

The plantar grasp is thought to be a vestigial reflex[5] yet it is still very much used in carrying. Grasping of the toes helps with gripping onto clothing or skin and this reflex is one that can also set off a chain reaction of other reflexes when elicited on-body. Its use is seen off-body too, during interactions with the world around them. It has been said that there is a strong link between standing and the reflex integrating[6] yet it's clearly present in early walking, aiding the baby's achievement of balance. The integration process is more obvious with this reflex as it requires greater amounts of pressure to elicit as the child begins to stand assisted through to walking unaided.

Babinski's - the splaying of the toes and sometimes- dorsiflexion of the big toe, which is elicited by stroking or rubbing of the side or sole of the foot – is a very useful reflex in carrying as it helps to stabilise the baby if their foot slips down. It works as a sort of brake. Interestingly, although it is elicited when a foot slides down the caregiver's body, the *act* of triggering it also elicits a flexion of the leg, which isn't really talked about. The flexion we see is the clinging adjustment reflex also being activated. The reaction triggered widens the surface area of the upper foot. When dorsiflexion of the big toe is also present it allows the baby to "hook" their toe onto the clothing/skin. In the lesser reaction (when not triggered to full extent) we see just a fanning of the toes, which is still helpful.

Stepping reflex/Clinging adjustment reflex (CAR)

As mentioned in my previous book, this is one of my favourites as it is a reflex which has already been discovered but is misunderstood. The CAR is widely known as the "stepping" reflex, and this is due to it being tested off-body, applying pressure to the soles of the feet by having a very young baby cling to the doctor's fingers with their hands, or supported around the upper body and letting their weight bear on their feet. This elicits a hugely exaggerated "stepping" forward action which – when viewed on-body – is actually a movement which helps a leg which has slipped down adjust itself back to a flexed state and grip tighter again.

When this reflex is elicited off-body the baby is able to step over obstacles. This may also be related to the lower extremity foot placing reflex – as the toes encounter an obstacle the same sort of reflex is elicited – it's just named differently. The term baby is also able to "walk" up an incline, which (as we will see in the next chapter) is especially useful on-body in a reclined position. However, even in newborns, the reflex is elicited by brushing motions on the side and sole of the foot, as mentioned in the Babinski's section. It's not exclusively a weight-bearing reflex. It's viewed as a stepping action because off-body we see a relaxing and lowering of the leg after the initial "step". This visually seems like a walking action when tested in the upright position; especially when coupled with the tendency of people to physically move the baby forwards when testing it. On-body, we tend not to see the leg lowering back down, because the foot is still in contact with something. Off-body the foot has nowhere to rest after activation, so the leg disengages once again.

The "stepping" pattern of the foot is heel-toe in term babies but starts out as toe strike in older babies starting to walk unaided. Viewing this reflex as "stepping" makes little sense, especially with the knowledge that heel-strike in independent walking develops later on. The heel-toe strike seen with this reflex serves the carrying process wonderfully, as it ensures easier access of full contact of the foot to the body they are on. The pattern is created by the dorsiflexion of the ankle as the leg hitches up. As the foot connects with the body it is the *ball* of the foot which makes first contact, due to the orientation of the surface they are "walking" on. This allows for further stimulation of reflexes, such as the plantar grasp, as the rest of the foot connects lower down. In the instances where the side of the foot connects first (as in hip carrying), the foot is still angled in a way which takes advantage of the heel being lower, being a stopping point. In fact, sometimes we will see at this angle the heel actually connecting with the body either first or exclusively – it all depends on how the baby is adducting the foot.

If the pattern were toe-heel then the foot would be angled with toes pointing downwards, which wouldn't help the carrying process due to the way in which the exposure of the ventral foot is affected at different angles. Feet tend to be at an angle of 90° or higher when connected to the body in active hip carrying, and at these angles more of the foot is available to make contact with the carrying person. It would also mean the foot is angled in a way where the heavier part (the heel, connected to the leg) is bearing less weight.

Another interesting thing about this reflex is that it's commonly reported that it integrates at around 2 months old, yet it is clearly seen on-body in babies up to around 6 months of age. Funnily enough, it's been found that if the reflex is regularly stimulated it doesn't disappear that early – it stays around for roughly 6 months, just as we see when it's elicited on-body![7] This revelation makes for interesting thought about how long nature intends reflexes to be around for. Do others stay for longer when regularly stimulated? This also adds confusion to the idea that reflexes need to be used to be integrated; that a certain amount of repetitions need to have been made. I believe that it's unlikely that regular stimulation makes a reflex stay for longer, as we know it's detrimental to development for them to be retained past a certain point. It appears to be more likely that there is some connection between the type of stimulation and how often it's used, as we know this reflex is still present well past the generally agreed "inhibition" age of 2 months.

Interestingly, though, there is some thought that it is weight which impedes the reflexive action, as it's been found that once the reflex appears to have integrated it can still be elicited when the lower body is submerged in water.[8] This finding gives weight to my belief that this is most definitely a carrying reflex and that there may be discrepancies about ages due to the usual method of testing to elicit it, which is through weight-bearing. As a newborn baby is brought to the caregiver's body, there may be some weight-bearing actions going on before improved tone is developed, but as they mature the nature of the stimuli is usually a brushing action of the leg and foot on-body. How, though, is it still elicited through weightbearing when babies regularly stimulated retain the reflexive action from this provocation, as found by Zelazo et al. in citation 7? As mentioned previously, maybe it has something to do with how often the reflex is elicited in a certain manner.

We know that integration happens if they are stimulated "enough" and that for integration to be possible there must be a "readiness" for it to happen. Zelazo et al. found that when the stepping reflex is regularly stimulated the number of steps produced by the baby increase. This is consistent with on-body clinging behaviours of seeing the clinging adjustment happen more in active carrying, especially when the baby is given regular opportunities to cling. This, combined with the presence of the reflex for longer than the usually accepted timeframe, indicates that repetition is not a way of integrating said reflex faster. It may also suggest that babies have an opportunity to improve the *quality* of their reflexive actions during this time period, potentially leading to better performance of the voluntary action to come.

This reflex is seen a lot in younger babies in skin-to-skin holding when active and alert, biological nurturing and the chest-to-chest hold. It's also seen in the shoulder hug but not as much. It really comes into its own in carrying when the baby moves to being carried on the hip. As babies cling reflexively to the side of the body, the movement of the caregiver will sometimes trigger slippage of their legs. This reflex helps them to readjust their position. As it integrates the caregiver may find they need to physically readjust the baby for a short while to encourage the adoption of voluntary actions, but this tends to be more so when active carrying hasn't been used previously and they're just now switching from passive carrying. If they had been carrying actively but do find a bit of a lag around this time, it can sometimes send a message to the caregiver that the baby is losing its ability to cling. It's here that a switch to passive holding sometimes occurs, due to the misunderstanding. Just as when babies learn other movement patterns and work towards physical developmental milestones (such as actively reaching out and grabbing hold of something), the carrying process also goes through refinement.

The learning of voluntary clinging adjustment appears to be linked closely to the integration of the reflexive squat action. Soon after the CAR integrates, voluntary disengagement from the carrying process appears in many babies and this is possible once the squat reflex is being integrated. The plantar reflexes set off a chain reaction which trigger the CAR, which in turn triggers the next reflex we will look at - the "clinging reflex".

Clinging reflex

This is a reflex of many parts - it is seen mostly with the thighs and knees, but it may additionally be elicited from the calves, ankles, feet and even toes, depending on the position the baby/child is in on the carrying person's body and how much sensory input they are receiving. It's a sort of umbrella term which includes all of the above reflexes as they can all work together. As you will find out later on in the book, clothes provide a spectrum of barriers to clinging behaviours, so we most often see thigh-clinging, then thigh and calf, then thigh, calf and ankle and so forth, depending on the layers involved.

This reflex is triggered by body contact on their inner thighs, eliciting a gripping/clinging action from the thighs and knees. In a hip carry, even if their legs are bare, they are unlikely to cling with anything other than their thighs and knees with the leg which is to the back of the caregiver's body. The dominant clinging leg is the front leg, which makes sense as when we walk we make forward motions, with the child needing to compensate more there for the displacement.

There may be exceptions to the dominant vs. non-dominant in some babies and children, as I have learned with my youngest. He has a right-side preference and eventually began compensating for his weaker left leg by using his right leg more in a left-side carry. I'm not sure if this was to do with leg length suddenly making this possible (he only began this compensation at 2.5 years old after a whole year of side preference) or a "lightbulb" moment for him. Whatever the case, he did this for a few weeks then it lessened greatly, and he rarely does it anymore. This could be to do with the fact that summer had arrived in all its glory and we were doing much more skin-to-clothing carrying outside and skin-to-skin in the house, which made working on the side-preference much easier.

This reflex is essential for babies to learn how to cling to their caregiver's hip/side. It is present from birth but begins to integrate as they begin to sit, alongside the spread-squat reflex. These two reflexes are linked together. This tends to mean a fairly small window of time for them to get used to this part of the body before they need to start learning to perform the action themselves if they aren't held here from the time they're developmentally able to. As with the CAR, we must remember that for many babies there is a transition phase where the caregiver may need to physically adjust their position as they learn the appropriate actions themselves. Again, this integration period can signal a switch to passive holds if the carrying person isn't aware of the potential learning curve the baby has come up against. Babies tend to learn very quickly, so if they're guided back into the clinging position they should soon learn to activate the correct muscle groups themselves. This is why reflexes don't just magically disappear overnight. If there's no time for them to learn and adjust to having to do the work, they would surely give up! The adjustments needed from the caregiver (if any) tend to be subtle unless they've not realised what is happening and the baby has started to get into passive mode.

Upper extremity proprioceptive placing reflex and foot placing reflex

The upper extremity reflex presents as a sharp flexion of the elbow then an extension of the shoulder and elbow, which follows immediately after. The wrist and fingers extend and abduct but may then follow through into adduction of the fingers, whilst the shoulder, elbow and wrist all stay extended. It's elicited by stimulation of the dorsal hand. This is again a stabilising action in carrying and will develop from the placing of a closed fist to a hand with extended fingers going forwards.

The foot reflex is elicited by stroking or rubbing against the dorsum and the response is a jerking up of the knee with forward motion of the leg, similar to the clinging adjustment reflex. In carrying, it's simply a way of eliciting a leg-righting reflex if the top of the foot is stimulated rather than the bottom.

Asymmetrical Tonic Neck Reflex (ATNR)

Also known as the "fencing reflex", this is elicited by baby's head turning to one side. The response is for the arm and leg to straighten on the side the head is turned to, and the arm and leg on the opposite side to flex. It's present for around the first 6 months[9] and in carrying this action has a stabilising effect for the baby. Just as with all the other carrying reflexes there is a spectrum of strength of the elicitation of the reflexes and this can affect the full response. For example, while the arm extension is triggered immediately when the head is turned, the leg won't necessarily extend right away – or at all sometimes. In fact, during the second month of life it's been found that the leg responses decrease.[10] Also, if we see the leg extension, it's common to see the leg draw back up soon afterwards.

On-body this makes sense. The main part of the body needing stabilisation initially is the head, seeing as it is very heavy on a young baby's body. The reflexive arm-straightening is similar to how we throw out an arm or two when we lose balance. If the head turning was a bit forceful then the leg will kick out too. This action can serve as a brake on-body, thus causing a chain-reaction of lower-body reflexes as the foot stabilises and the leg then hitches up again. The fact that the reflex begins to inhibit when babies have head and neck control is telling. Babies heads are heavy and having a stabilising reaction helps to counter the effects of head movement. The start of learning the voluntary actions at this point is perfect timing, and still allows for reflexive correction of the less-stable lower body.

Once a baby is carried on the hip any lingering ATNR may aid carrying by the flexion of the arm connected to the caregiver's arm enabling better bracing whilst the extended arm on the side they're looking away from the caregiver aids balance. An absence of the leg extension means they're still able to cling with the lower body, but even if this is still mildly present it makes for a slight "braking" action, which will then stimulate the reflexes of the foot, enabling the leg to right itself.

Tonic Labyrinthine Reflex (TLR)

This beautiful reflex is triggered by the head moving forwards or backwards. As a vestibular reflex, one of its purposes is to maintain balance and posture. It's such an interesting one as it's directly linked to spinal development, causing the spine to either straighten out or curl in. Off-body it is elicited to either extreme – when tested in the supine position, the head hanging behind the midline causes a reaction of an arching back, extended legs which are brought together, and pointed toes. When tested in the prone position the head is left hanging in front of the midline and the extremities curl up. On-body we see a similar reaction, but, due to other reflexes being activated at the same time, it can sometimes look a little different. We're also able to observe what happens when it's elicited in a more gentle and natural way.

This reflex tends to be in action most when baby is in an active-alert state. At this time, they are practicing head and neck control, and this triggers something in-between the extremes – a gently straightened spine. This is why even very young babies appear sit fairly upright on-body or in a sling/carrier when they're not pulled in tightly against the body. A spine in a straightened position is not something to be feared in a young baby using this movement. In the active-alert state they are exercising, developing their muscles, creating a strong framework to stabilise the spine, thus enabling control. They need to experience this movement to aid spinal development before they're able to make certain voluntary movements.

As the baby tires (or wobbles in its immaturity) their head comes forward and this triggers the whole-body flexion, which is well-known to be a position of rest which conserves energy.[11] If they aren't being supported well and a hand isn't at the ready to catch a head falling backwards then the extension reaction will occur. However, an extension of the legs may not be observed, or only a partial one, if the lower body reflexes are also being activated. It's all dependent on the level of contact with the carrying person's body and what the infant's legs and feet are doing at the time of the head dropping behind the midline. This is a handy adaptation as to fully straighten the legs can be extremely unhelpful in carrying, but the shift in balance from the relatively heavy head of the baby tipping backwards (which will feel even heavier, with it being even further away from the caregiver's line of gravity) and the arching back should alert the caregiver quickly to a need for more support and adjustment. This all tends to happen relatively quickly of course, and an attentive caregiver will rarely find the infant's head tipping right back, as they catch it as it moves.

Going forwards, once the flexion (*f*) element of this reflex is integrated the general movement of the spine will become more voluntary but will also be affected by things like the baby relaxing into the caregiver (the spine naturally curves slightly). When they push away they will still trigger the extension (*e*) part of the reflex, as this hangs around for around 3 years. Once TLR (*f*) has integrated another reflex emerges, which is the symmetrical tonic neck reflex (STNR). This reflex is not present at birth and for this reason it is classed as a postural reflex rather than a primitive one. The development enables a change in how the extremities respond to head motion. When the head comes forward, flexion is produced in the upper extremities and extension in the lower extremities. The opposite reaction occurs when the head is extended. This action is essential for adopting the position in which crawling will occur.

Its significance in carrying is that the reactions are elicited in the extreme – head lifted up in the crawling position (or tipped right back on the vertical plane) or head brought down, which does not happen in carrying. As the clinging reflex disappears (which aided the inhibition of leg extension when TLR (e) was elicited), STNR enables reflexive flexion if the baby decides to fling their head backwards. The reflex doesn't interfere with regular carrying as it's not triggered during the normal head movements seen.

Moro and startle reflexes

The Moro reflex (sometimes also called the "startle" reflex) is elicited when babies feel unbalanced or in response to loud noises. When tested off-body tends to present as a sudden extension and spreading of limbs, freezing, crying then drawing back into flexion. This is because the person testing is trying to elicit the reaction in extreme. On-body this is a much gentler response during carrying, and very rarely will we see the full reaction; it would be a hindrance for the baby to fully spread their limbs and extend them. In fact, if the baby is engaging with their hands as well we can expect to see a just the second half of the reaction.[12] This would be especially useful for making sure the cry response isn't elicited from the baby if the reflex was triggered by the caregiver fleeing from a dangerous situation. This reflex should not be seen past 5 months of age.[13]

The way Moro tends to work on-body is as a stabilisation action from motion, and is subtler than one would expect, meaning it's not always noticed by the untrained eye. It seems to be that it mainly helps by eliciting flexion as a baby wobbles and this of course makes sense because the baby will usually be fairly well supported body-wise, but the vestibular sensory input will send messages of some instability. The extension (however fully elicited) is helpful in that it enables a sort of reaching action and reflexively opens the palms which makes it possible for the baby to "grab" a hold upon the next part of the reaction – adduction.

As the baby gains good head and neck control this reflex begins to integrate. Integration of this reflex relies heavily on the fear paralysis reflex having already been integrated, which should have happened by birth.[14] Moro is replaced by the startle reflex/response, which stays for life. This reflexive action sees just the flexion phase of the Moro reflex, and is obviously better suited to now-actively clinging baby. Moro is present during the phase of carrying where babies are naturally more supported by the carrying person.

As you can imagine, eliciting either of these reflexes to the extreme is not advisable - it releases adrenaline and cortisol as well as increasing heartrate and blood pressure. The subtler responses are from a less stressful trigger, and the extreme reactions appear to be linked to survival instincts. The full responses will not be a regular occurrence in carrying.

Palmar Grasp

The grasping reflex of the hand is activated by pressure or light stroking of the palm and is deactivated by stroking the side of the palm near the little finger. It's obviously very useful for enabling the grasping of objects as babies become more of the world around them but have yet to learn the motor control needed for voluntary grabbing. In carrying it's used for grabbing onto the caregiver's clothes, skin and/or hair, as well as aiding sensory feedback. It's seen as a stabilising action of the upper body and is usually coupled with a connect of their forearm to the caregiver's body.

It can be reversed by stroking the lateral side of the palm, which elicits an uncurling of the fingers from the little finger inwards. In very young babies a normal response is flexion at the elbow if any weight is borne by them, which helps keep the immaturely developed infant both attached and close to the caregiver's body. The reflex develops further at somewhere around 4-8 weeks of age with an appearance of the beginnings of individual finger reflexive movement.[15] Stimulation of the thumb and forefinger will elicit flexion of just these digits. After a time, each of the rest of the fingers will become subject to isolated reflexive movement by stimulation of the palm by the base of each finger.

Grasping is particularly useful as the baby becomes more aware of the world around them and takes interest in interacting with it during carrying. A reflexive hold of the caregiver enables better balance and stability of their upper body as they look and move around. As the reflex begins to integrate they are learning to voluntarily grasp objects and gaining smoother and more coordinated control of opening and closing their hands. This translates well to carrying as they make more sense of the carrying environment (caregiver's body/clothing) and how to interact with it in voluntary ways.

Cremasteric reflex

This superficial reflex is elicited from light stroking or pressure of the inner thigh. The response is that the testicles are drawn up into the body by the cremaster muscle. It is a lifelong reflex, not a primitive one, and its official use is thought to be to help with heating and cooling, as well as protecting the testicles. Funnily enough it's also elicited in carrying, making it yet another carrying reflex! The pressure from the inner thighs activates the reflex, protecting the testicles when clinging (and in some passive carries, depending on where the thighs are), with them descending again after the pressure is discontinued.

As with every new revelation when it comes to carrying, the benefits of this reflex are clear. Of course, the testicles would need protection from the pressure of sitting on-body. In addition to this, when a baby or child is carried skin to skin with a tilted pelvis the penis and scrotum are able to gather upwards in the "pocket" created between the caregiver's body and theirs. This provides some safeguarding for the testicles, but it's the addition of this reflex which ensures complete protection.

Since my discovery of this reflex also being a carrying one, I've understood Isaac's occasional cries of discomfort. These days if he says "ow!" and wriggles uncomfortably he will then spontaneously let me know or I can ask him if I've hurt his testicles. I then readjust his position, making sure his inner thighs make contact with my body before I lower him into a seated position. This has always resolved the problem.

Sleeping

As we know, some reflexes are much weaker when babies are sleeping, and sometimes they won't be elicited at all. This in itself is a plausible reason for humans creating baby-carrying devices. If babies assist in the carrying process but this diminishes quite a bit – or disappears altogether – when they sleep, what are you going to do? You're unlikely to pop them down somewhere to sleep alone, unprotected from predators. Even the fact that close bodily contact regulates the sleep cycle[16] (albeit shown via skin-to-skin contact) is reason enough to keep them on-body, rather than have them sleeping for shorter and more unsettled periods. The increase in perceived weight when they are participating less (or not at all) is rather noticeable, even in the early weeks. Muscle tone also decreases greatly, so with these things in mind it becomes clearer that in-arms carrying may not be particularly suited to when they're sleeping (unless that's what the caregiver chooses to do, of course).

Reflexive carrying in action

So, reflexes tend to work in teams and it's not that common in normal, natural, on-body movement that just one is elicited at a time. When reflexes are working together they can also look different to how they display off-body in a completely different position. The sooner we move away from separating reflexes and move towards observing them in groups which come together to create smooth, coordinated movement, the more we will understand when watching them in action on-body. Separating them is useful for studying the different parts of the reactions but not for understanding how the reflexive carrying process works *as a whole*. In fact, I would even go so far as to say that what we call individual reflexes are actually separate stages of a bigger reflex as a whole. Let me explain why.

If we observe what happens with the lower body in the earlier stages of reflexive carrying, we see lots of things happening. Legs in contact with the body, some disconnect maybe, feet brushing the body, leg movements, foot movements, toe movements. We can single out certain movements such as the clinging adjustment reflex, and notice different reflexes being triggered whilst the bigger movement is taking place, and as a result of the ending of the reaction. For example, in CAR we may see the plantar reflexes, placing reflex as well as the clinging response. Reflexes tend to build upon each other to create these more complex movement patterns.

If we observe a young baby who is alert and active in a shoulder hold, who has some neck strength but is still working towards full control, the following is a common sequence of events of the upper body seen in many babies. The infant may rest its outer arm on the caregiver's shoulder and inner arm against their chest. They're sat at a 90° angle and they use their arms to support their upper body whilst looking around. Their interest in the environment surrounding them causes them to gaze, which, depending on what they're looking at (and sometimes whether the carrying person is walking/moving), may result in them moving their head. This motion of the head creates movement in the upper spine if it's not abrupt and engages muscles of the upper back. If they move their head to the side we may see subtle adjustments or more pronounced, again depending on the force of the movement. The movements may be jerky at times and set off Moro whilst TLR in extension is also being triggered – TLR moves the spine straighter or into a slight arch and Moro may cause a startle reaction before the second half (flexion) occurs.

If we think more about TLR, and factor ATNR in as well (both triggered by movement of the head), we discover even more happening. These cause reactions through the spine and can also trigger leg movement when elicited in full. However, the way babies tend to sit on-body can counter the lower-body reaction, as reflexes of the lower body are stimulated at the same time. For example, say the reflex trigger was quite strong and their leg/s started to straighten – on-body this is likely to mean that the leg and foot would push/brush down the caregiver's body, which would then trigger stabilising reflexes such as Babinski's, plantar grasp, clinging adjustment reflex and so forth.

Now let's look to the older baby who is still in the stages of reflexive carrying but is learning to cling to the side of the caregiver. Here we have less of the upper-body reflexes still in play but all of the lower body and some of the vestibular responses are in full swing. The clinging reflex is now being actively utilised, and with this stays the activation of the plantar reflexes and clinging adjustment reflex seen in front carrying. Every time a leg slips a little (or a lot) we see a chain reaction of lower body responses. Every time they move sharply we see upper-body stabilisation. Though reflexive upper body movements have been replaced by voluntary actions, subconscious movement still occurs. The baby doesn't have to think about every step of the process even though they may be able to voluntarily control some of it. As you can see, reflexes truly are designed to work together, and when they do we see distinctive movement patterns emerge.

A matter of repetition?

There is compelling evidence that when a reflex is stimulated regularly it results in an earlier emergence of the voluntary action.[17] This may sound at odds with the previous discussion about simply repeating an action over and over not meaning that a reflex will integrate earlier. The key is, of course, the factor of *readiness* for the integration to occur. Once a baby is in the zone of readiness, then repeated actions may have an impact on the achievement of said developmental milestone. Bower and Zelazo both conducted research in 1976 (separate studies) focusing on different reflexes to see whether regular elicitation had any impact on the emergence of the skill it was related to. Both found that it did. This is especially interesting as it seems clear that if the body is allowed to move in all the ways it was meant to and engage in the activities it was designed for, there could emerge a new "normal" for developmental milestones. Motor development relies on repetitive movement. Another interesting finding from the one of the studies was that even small amounts of practice of the movements caused significant effects, indicating that there seems to be some sort of "memorising" of the reflexive movement patterns by the brain as development occurs. This is in alignment with my unofficial findings that clinging ability does not rely on huge amounts of practice – active carrying just needs to be offered and practiced on a general basis.

Going back to the fact the "stepping" reflex was present much longer, and became stronger before it gradually integrated, there is something else to consider from that study. The way they tested this was to have a control group (no exercise), a passive exercise group (laying supine, with caregiver moving their legs up and down) and an active exercise group (supported upright in a loadbearing posture). From 3 weeks old the no exercise group rapidly lost the reflex. The passive exercise group showed a small increase before dropping off, but the active group showed a sharp increase, followed by an even bigger one from 6 weeks of age. This suggests there is much more participation and muscle stimulation from eliciting a reflex than from the caregiver moving their legs. The significance of this is that holding babies in static positions or doing the work for them does not produce anywhere near the same effects as allowing them the freedom of movement and opportunity to complete the actions with the natural sensory stimulation required to do so. This suggests potential implications for the baby's normal physical development when passive carrying and/or babywearing is used as a replacement for active carrying.

More research was conducted in 1993[18] replicating the initial study, with an addition of a new test – practicing sitting. This time there were 4 groups. There was the control group (no exercise), the stepping group, the sitting group and a stepping plus sitting group. In each group they were tested every 2 weeks to see how much they were stepping and for how long they could hold a sitting posture unaided (but supported at the thighs and buttocks). Again, they found a distinctive positive trend in development in the active groups. I mention this as – although the sitting test was off-body – this suggests we may find a link between greater postural control in babies who are carried actively, allowing for both reflexive movements and less restriction for the upper body. It would be extremely interesting and beneficial for research of this nature to be conducted going forwards.

Overall, the prevalence of the reflex beyond the generally agreed norm, along with the fine-tuning of said reflex, seems to be in line with my unofficial observance of clinging adjustment in the infants whose families I have worked with and those I've observed in relevant situations. From what I've seen, stimulation of the reflex on-body corresponds with a longer duration of reflexive elicitation, carrying on through the early stages of hip carrying, until the squat and clinging reflexes are integrated. Once this occurs we tend to see voluntary clinging adjustment, or, in the case of the child who hasn't been offered active carrying opportunities, a loss of positional adjustment on-body.

The premature infant

As primitive reflexes develop in the womb, being born too early can mean some of the reflexes have not yet developed. Premature babies are at greater risk of developmental delay and it's crucial that they receive appropriate sensory stimulation to encourage the development and integration of reflexes. Of course, they are not receiving this in an incubator. Dr. Harold Blomberg of Rhythmic Movement Training suggests that the gestational parent carrying the baby on their chest provides the infant with similar sensory stimulation to the womb. Of course, there will be a timeline of progression to this based on how early a baby was born and whether or not it is possible for them to be carried. However, reflexes can be stimulated on-body in other positions, such as during kangaroo care, meaning that the benefits of carrying for the reflexes need not wait. It's also been found that – when their age is adjusted to account for their prematurity – there is no difference in the amount of time the primitive reflexes are present.[19] However, it's also been shown that premature babies are at a greater risk for developmental delays[20] so something to explore further would be why this is so.

Another point to be aware of is that premature infants are less likely to spend much time in an active-alert state to begin with.[21] This means that the normal stimulation of primitive reflexes is likely to be less than that of a term baby. Although different reflexes may be elicited during varying states of alertness, spontaneous reflexive movement relies on the infant being active enough to move their body in specific ways. This is interesting because a diminished capacity for being in an active-alert state would help with conserving the vulnerable baby's energy but may also be a potential avenue of exploration to look for answers to the question of why they may be at greater risk for developmental delay.

Chapter 1.4

Breastfeeding and Carrying

A big question to ask is: what role does infant feeding play in the carrying process? Well, it appears breastfeeding may be a major link between being born and being carried and may aid the carrying process in the weeks ahead. In this chapter I will discuss breastfeeding as it is the biological norm and we're seeking to better define the biological baseline for carrying. There are specific links between babies searching out the breast and behaviours in early carrying. The information discussed obviously raises questions around what may be done to "fill in the gaps" when a baby is bottle fed and may provide some ideas going forward.

The biologically normal habitat or environment for a newborn and young baby is the gestational parent's body. We know this because of many things, including:

- The infant's inability to fend for themselves

- Their limited ability to get from point A to B

- Their inbuilt need for human touch

- The make-up of human milk, requiring little-and-often feeding

- The biological processes which make the gestational parent lactate, making them the primary caregiver from an evolutionary point of view

The concept of the "fourth trimester" is becoming more well-known and accepted these days, and in basic terms means that from birth to 3 months of age the caregiver/s facilitate an environment which creates a "womb on the outside". This means plenty of bodily contact (via carrying and babywearing, if they choose to use slings/carriers), gentle movement, readily available nutrition and so forth, to enable a smooth transition from womb to world. I believe that the fourth trimester may last a bit longer than 3 months and the end may actually be signalled by readiness to transition to the hip, as this heralds both a new way of interacting with the world around them and the reaching of the required physical development to do so. This will obviously differ between individuals but an average of around 4 months appears appropriate.

In the beginning the babies' senses are less developed, especially their sight. Being close to the caregiver/s promotes a sense of security during this time and enables them to learn about this new environment and get to know the outside of their first home before getting used to other parts of the world around them. It's a bit like if you had lived all your life underground and suddenly were ejected into the full outside world. Imagine the overwhelm and sensory overload! You would likely be in need of a gradual transition, first getting used to a room indoors, then exploring other rooms in the house, maybe the garden next, and so forth. The baby is developing rapidly and with the sensory and physical leaps taking place they're soon able to start interacting with the world in new ways. Having this facilitated by being in the arms or on the body of the trusted caregiver gives security and offers different learning opportunities.

The initial exploration for the newborn is finding its new source of food now they're no longer nourished by the placenta. It's been suggested that in the first few days, before the colostrum changes to breastmilk, that hunger is not a primary factor for feeding.[1] This does make sense if we think about the fact that they spend very little time awake during this time. If hunger was present every time they were awake, and suckling eventually sends them back to sleep, where is the connection to the caregiver? Where is the learning experience? Hunger hinders concentration and drives us to seek out food, not comfort.

Breast crawl

The phenomenon of the breast crawl has been greatly researched.[2] The crawl occurs when the newly born baby is placed on the gestational parents abdomen and is left to independently seek out the breast. This involves the activation of primitive reflexes such as the crawling reflex (which is similar to the CAR), rooting, grasping and sucking.

Breast secretions (not the milk) are said to contain certain substances that are also in amniotic fluid and this is a reason why washing the infant is discouraged immediately after birth if caregiver's want their baby to experience the breast crawl.[3] This link to the womb to seek the breast out through a familiar smell is a very useful adaptation. Not only does it enable the baby to find its source of nutrition, but it also sets in motion the activation of reflexes involved in motion, which enable them to reflexively "crawl" up the caregiver's body. The breast crawl is ideally facilitated by the gestational parent being horizontal as it means the newborn is able to propel themselves forward more easily than if they were in a reclined position.

As we know, babies are designed to use specific reflexes at specific times, such as during the birth process. The fact babies are able to independently seek out the breast immediately after birth raises the question of whether they're also expected on a biological level to complete the breast crawl post-birth. Whilst this may be debateable, the fact that reflexes are so easily stimulated skin-to-skin suggests that there's a possibility that there's a built-in expectation for the stimulation of them at this time. Stimulation of the birthing parent's abdominal area is known to help uterine contractions[4] which controls the post-partum bleeding and aids in the shrinkage of the uterus. Having the baby on-body, skin-to-skin, means no barriers for sensory input therefore the crawling/climbing actions are better stimulated.

If we go with the theory of hunger being the driving factor once the milk comes in, the baby now knows that their food is coming from this body, and where on the body to get it from. They're familiar with the smells and feel of the caregiver's body, and will now learn the smell of breastmilk, which will let them know where their food is when they're near the breasts. Breastfeeding is a key, biological, factor in making sure the caregiver is constantly available and provides regular body contact. It also seems to have a pivotal role in ensuring the newborn gets to know its carrying environment.

Biological nurturing

As we know, most reflexes are present from before birth, appearing in utero at different gestational ages. Some play a big role in the birth process, enabling the baby to reflexively navigate the birth canal, making vaginal birth possible. Others are present during feeding and the "breast crawl". Suzanne Colson's research into the "biological nurturing" position gives us a great insight into the links between feeding position and carrying.[5]

The biological nurturing position is when the caregiver is slightly reclined, making available all of their torso for the baby to connect with and explore. The baby is placed upright on the caregiver and in a way that the front of their body is on the torso. When this happens a host of reflexes are activated. In the same vein, when a baby is placed on the horizontal caregiver in the same way, they are able to seek out the breast due to the activation of a host of reflexes, which enables them to physically move themselves up to the breast through the breast crawl movement. Biological nurturing is possible for most term babies immediately after birth. Many of the movements observed in these positions are the same as those seen when carrying a newborn chest-to-chest and in the shoulder hug. Just like the breast crawl, in BN the actions of the feet and legs, for example, are of a crawling motion when the caregiver is reclined or lying flat.

The biological nurturing concept has many similarities to newborn carrying principles. Colson recognised the need for what she termed "frontal feeding" positioning. This enables the baby to not just be tummy-to-caregiver – it allows their chest, pubic bone, thighs, calves and feet to all have full contact with them or the surrounding environment. She found that in this position babies were able to latch on more easily and approach the breast independently, that the reflexes triggered appeared to be smoother and more coordinated, and they required no holding or physical pressure from the caregiver. Of course, the difference with BN is the hands-off approach, which isn't possible in carrying due to the fact that our bodies are upright when we are walking around. Babies obviously need some support when we are upright and moving, but when we are sat down and reclined in the biological nurturing position they are supported enough by the angle they are at, and the caregiver can easily offer support here and there if needed.

This does, however, parallel the in-arms approach of ensuring the baby has as much freedom of movement as their stage of physical development and state of alertness allow. Following their lead and not going overboard with support when they are active and alert, enables them to both strengthen muscles and activate primitive carrying reflexes. In biological nurturing it was found that allowing babies this optimal learning environment ensured earlier coordination and conditioning of the reflexes.

It's fascinating that feeding reflexes trigger a chain reaction of other reflexes, just as we see in carrying. As we found out in the previous chapter we see a domino effect of reflexes being set off when the plantar reflexes are activated, ensuring the baby is able to adjust position and engage in the carrying process. In the biological nurturing position, they found that the plantar grasp and Babinski's sign were linked to the reflexes of the mouth, and that when babies fell asleep they could check if they had finished feeding by stroking their foot.

Postpartum recovery

Now, going backwards a bit, let's think about what is biologically normal for gestational parents. Birth is an incredible process, but even if they have had the most normal and undisturbed birth possible there is a recovery period. Take it from someone who's had 4 normal births including 2 undisturbed ones, no matter how "back to normal" you feel and on a complete oxytocin high, it doesn't magically take away the fact your body has *laboured,* expelled a human from your body, the uterus needs to go back down to normal size, organs need to return to their normal positions, hormones need to return to pre-pregnancy levels and so forth. What would be normal for the gestational parent is for them to have a period of rest in the weeks following the birth. In England this was traditionally 6 weeks in most recent history. In other cultures, it's between a month and 6 weeks. Let that sit with you. 4-6 *weeks*. This rest period tends to begin with bed-rest, then smaller periods of movement, building up to reintegration of normal life. So, if the caregiver is recovering, would much carrying be done in this initial period? It's plausible to hypothesise that there wouldn't be much at all. There would be a lot of rest for them and the baby in the first days, and some small amount of carrying, maybe by another parent, grandparent and other relatives to begin with, but unlikely for long periods of time or of any distance.

This raises the question of what would happen about encouraging the coordination of the "carrying reflexes" during this time. Well, if we go back to the biological nurturing position and the breast crawl and think of them being used in this rest period, thinking about the similarity of these movements (minus the seeking out of the breast), it becomes clear that they may well be physiologically linked to the carrying process. Having the reflexes stimulated in this way – on horizontal and diagonal planes – would mean the baby would be practicing lifting its head, cycling its legs, propelling itself forward with its feet and legs, grasping, as well as getting to know the "landscape" of their caregiver's body. This would also be very useful for the caregiver, for getting to know their baby's movements and behaviours, feeling how they fit to their body and adjusting to how to support them on-body. With an awareness of what these actions mean, and that many are triggered in carrying, the caregiver is better placed to recognise and work with these behaviours on the vertical plane.

Once the caregiver has recovered enough to be up and about again, and carrying lots, how does the role of biological nurturing shift? I believe that both practices work well side-by-side, each supporting the other; that breastfeeding supports carrying and vice-versa. In the biological nurturing position, the baby is able to practice movements which are less conducive to being upright at this stage of development, such as raising their shoulders and working their upper torso. This is also a lovely way of spending time connecting with each other - especially if skin-to-skin – whilst doing something similar to "tummy time" (without a hard floor below their face).

I believe that breastfeeding has this additional role of being a link between the baby being birthed and regular carrying, and that biological nurturing can enhance this by providing a beautifully balanced environment for both the baby and caregiver to learn carrying behaviours. In this semi-reclined position, the whole of the torso is available to the baby. The angle allows for easy observation of the baby by glancing down rather than having to bend at the neck and the fact the caregiver is stationary means that it's much easier to direct their full attention to the baby and notice what they are doing. Even if they are limited to only a few days' rest and do not continue using the position regularly (or at all), I believe this initial time can help with the subsequent carrying process. As Colson points out, it doesn't usually take long for this coordination of feeding reflexes to happen so even just using the first few days before the milk comes in, when baby is less motivated by hunger in feeding, may be quite useful.

Feeding whilst carrying

As the baby develops further, the caregiver may decide to feed them on the go. This tends to be done in a cradle position in-arms or in a reclined seated sideways or upright position in a sling or carrier. It's interesting to look at the use of an upright position when feeding in-arms. The position the baby would be in on their body is off-centre and lower down than the shoulder hug. The position of their legs would be at the lower ribs or waist, depending on their height, and the position of the nipple means that they will likely be slightly on the hip too with their outer leg. Supporting them in a mainly passive hold by supporting the bum but keeping the legs free of support means they are able to rest to nurse but their legs are free to make the movements seen when feeding in a reclined or flat position, body to body. This position is obviously a bit awkward when their muscular development isn't very advanced. When it is, their weight tends to be an issue as it's not possible to support them at a low enough position with the forearm without compromising it. This makes suggests to me that the cradle position is designed for feeding on the go, especially before the baby is able to feed from an active hip carry. Our arms are designed to hold and support young babies with our forearms. The way a baby's head is supported in the crook of the arm positions them, so their chin is away from their chest and the hand is in the right place to support their buttocks. The height they're at and where their head is makes supporting feeding in this position easier.

The cradle carry is, of course, passive. This, along with the "passiveness" of the biological nurturing position – in the sense that it takes away the need for active clinging – lends support to the idea that babies are not primed to be clingers when feeding. It also appears that feeding is something we're expected to facilitate in a position of rest. Feeding on the go is possible but having to do most of the work supporting them is a drain on our energy. There are, however, findings showing that the palmar grasp reflex is stronger when a baby is suckling, which we touched on in the previous chapter. This may indicate some participation is expected at a primal level.

The babywearing connection

In a 2012 study in Italy it was found that there was a link between using baby carriers for at least 1 hour a day and increased breastfeeding duration rates in term babies.[6] This is obviously both wonderful for encouraging breastfeeding as well as keeping babies close. This may suggest that spending time on-body could help with sustaining breastfeeding. If being held statically in a sling or carrier has been shown to improve breastfeeding duration, what would we see if in-arms carrying were studied?

If we think about how babywearing induces stillness and calm in many babies, and how carrying in-arms when babies are active/alert stimulates them, it would be interesting to see how the caregiver noticing feeding cues is affected. Observations in general of caregivers responding to babies in slings and in-arms has shown myself and others that there is a difference in communication in each setting. If nothing else, in-arms carrying keeps babies close in ways comparable to babywearing, so it would be surprising if holding and carrying them negatively impacted breastfeeding rates.

PART II

Chapter 2.1

Participation in Carrying – Where Does it Begin?

It's easy to think of carrying as beginning when the baby or child is clinging to our body or sat in our arms, but the reality is that participation begins long before this. The carrying process begins at different points depending on who is initiating it. Whatever the individual starting point, it all starts with an expectation that they are about to be carried – whether that be at the point they're picked up or sometime before.

As research has shown us, anticipatory responses to the caregiver initiating carrying can begin as early as the point the caregiver begins to talk to them.[1] In the very beginning, with a newborn baby, the initial factor appears to be the voice of the caregiver. As the baby's vision gets better and they become even more self-aware, how and when they make anticipatory responses – both reflexively and voluntarily – changes.

Previously, research had shown that when a baby's attention is directed at another object they are able to generate direction-specific postural adjustments from around 1 month old.[2] As they approach 3 months of age the number of muscles being used in these adjustments decreases, and this has been suggested to be related to developmental transition of postural control. These increase after the transitional period. Another period of transition was found at around 6 months of age, when adaptive, secondary control emerged. From 9-10 months old subtle adaption of the degree of muscle contraction emerged and at 13-14 months postural adjustments were present. The new research sought to define a timeline based on anticipatory responses in relation to another person's actions towards them.

This sort of information is incredibly useful as knowing various stages of transition can help us to understand how postural adjustments develop and emerge, and how this is linked to picking babies up. It also helps with understanding some of the ways in which participation in carrying develops and adapts as the baby matures.

In this chapter we're going to look at both triggers – attention directed to self and to another object or person – as the process of being picked up involves several types of action, attention and behaviours. The arguments in the study are for prior planning for motor behaviour which requires potential awareness and that postural adjustments are more effective if they happen before the destabilising event than during. It appears, though, that it requires a balance of a combination of responses. For example, there's no way for the baby to be 100% sure of the exact way a caregiver is going to move when they begin each stage of the process of picking them up. Some postural adjustments *must* be in reaction to destabilising events. The idea that the event is smoother and more enjoyable for baby and caregiver is still valid, but it doesn't rely solely on anticipatory responses.

Although it was shown that there was no difference in anticipatory adjustments when a baby had higher or lower neck control, the study was focused on the pre-lifting stages. We would expect to see differences based on stage of physical development from the lifting stage onwards as there's the issue of displacement, vestibular stimulation and body-righting from this point. The earlier stages rely on hearing, sight and touch. The researchers believed that at 2 months old the adjustments happened "too early" as there were less adjustments in the chat phase at 3 and 4 months of age. We'll explore this further in the "chat" section.

The researchers postulate that the approach stage was a bigger trigger for postural adjustment than contact, but this needs further investigation (e.g. what happens when approached but not picked up?) as these are two different sensory triggers. Touch is a key stimulator, as we see in lots of primitive reflexes. It would be very surprising if it wasn't a major trigger in postural adjustments. The fact that the previous stage's responses didn't stop or change when the contact stage was entered shows us at that it's at least likely to be a natural follow-on step.

Approach is an interesting stage as it showed the highest levels of response. However, it was shown that gazing towards the hands may be a distraction rather than a help, which throws the arms-out idea as being a key factor in anticipating being carried into question. Is the reaction purely to do with the caregiver coming closer to them or is it linked to an expectation of being carried? What is it about the approach of a caregiver which calms movement? Is it the expectation of being carried, or merely the closer proximity to them?

Initiating carrying can involve up to 7 steps: communication from baby, chat/communication from caregiver, approach, initial contact, lifting, bodily approach and bodily contact. We'll find out what happens from both the baby's and caregiver's perspective, as well as how each stage links to the next. Anticipatory responses vary from baby to baby and will also differ based on the caregiver's "routine" when it comes to picking them up. For example, if they explain to the baby in some way that they are going to pick them up (e.g. "I'm going to pick you up now" or "Do you want to come up?"), they are more likely to associate this phrase or intonation with what comes next. If no direct "chat" is used, then it's unlikely that the occasional instance of talking to the baby before picking them up will be an indicator to the baby of what is about to happen.

As you can imagine, the anticipatory responses also vary based on who is doing the carrying. It's interesting how most babies have a primary caregiver. It makes a lot of sense for the baby to focus initially on this one relationship, learning the caregiver's smell, touch, body and so forth. Imagine the additional workload of having many people caring for them in equal amounts, having to learn and adapt to them all. Of course, for many babies, other people interact with and care for them too, but there is less of a *need* to work well together when they're not the person spending the most time with them.

This is not to say that having two caregivers sharing care equally is "bad" though. I believe this works well too, as a lot of principles are easily transferred between caregivers, it's just applying them to a different "environment" (person). Babies are also incredible skilled at adapting to environments and situations. It's just to say that from a biological baseline having a primary caregiver makes for more consistency and clearer expectations. Throughout history, gestational parents would be the people most likely to take on primary care duties and be the source of nutrition. Babies would be "tethered" to them in this way, so it comes as no surprise that nature would also include built-in mechanisms to make carrying easier for the person doing the most of it.

Communication from baby

If the baby/child is initiating the carrying process, it begins with some sort of communication from them. This may be verbal, by striking up conversation with their caregiver (even from a very young age), alerting them to a need with reflexive or voluntary sounds or fussing/crying. It may be visual, such as locking eyes with their caregiver, or physical, by making certain body movements. Usually, though, it seems to be a combination of 2 or more of these.

Babies are born with the ability to make reflexive sounds and these enable them to verbally communicate with their caregiver/s. Priscilla Dunstan discovered 5 sound reflexes which are heard from babies all around the world up until around 3 months of age, when the reflexive sounds begin to be replaced by chatter/babbling. These reflexive sounds have been named collectively as "Dunstan Baby Language".[3] The sounds and their meanings are as follows:

- Neh – "I'm hungry"

- Owh – "I'm sleepy"

- Heh – "I'm experiencing discomfort"

- Eairh – "I have lower gas"

- Eh – "I need to be burped"

As you can see, all these sounds are communicating a need, and in most instances would be a signal to the caregiver to pick them up to meet that need. Of course, not all caregivers will instinctively understand these sounds – especially in the early days of new parenthood – or even recognise that it's possible for them to mean something. However, for the sake of understanding the steps of participation in being picked up, we're going to take the approach that the fictional caregiver does understand.

Although this communication begins reflexively, babies will begin to associate their verbal communication with the appearance of a caregiver (if said person is attentive), and so they develop that communication further as their awareness and control of their voice and body grows. They will also begin to learn that if their initial attempts at communicating their needs aren't heard then crying (the last resort) may bring prompt attention. At this young age the verbal communication is coupled with reflexive movement, such as kicking their legs. Some of these movements may give the caregiver a clear signal of a need, such as the sharp drawing up and straightening of the legs when they have wind.

We know that caregiver's responding appropriately and in a timely manner to the baby's communication of their needs are laying down a foundation for secure attachment.[4] Focus tends to be on responding promptly to the cries of a baby, yet crying tends to be the last resort in a long chain of communication from them. The additional layer of communication – verbal or otherwise – in the process of being picked up offers yet another opportunity to foster healthy attachment.

As voluntary language begins to develop we notice cooing and vowel sounds. This develops further into babbling, and consonant sounds emerge such as "mmm" and "ppp". The significance of language development is that it helps caregivers distinguish different needs due to the type of vocalisation. Also, distinctive sounds and cries may be used to alert the caregiver to a need for being picked up. As development furthers they may begin to communicate their want or need to be held by reaching out with their arms, learning the word "up" or similar, using baby sign language or physical communication such as going up to the caregiver and tapping them. This obviously evolves over time to their chosen word, phrase or other method of communication.

Something interesting I've seen with Isaac in his toddler years is that he's developed two distinctive ways of asking to be carried. One is because he wants the physical movement and interaction and the other is when he has an emotional need which must be met. He says, "Hold me!" when he just wants to come up. When he needs comforting he says, "I need to hold you!". This is eye-opening for me. Obviously, it's usually easy to tell whether he simply wants to be carried for the interaction or because he's upset and so forth. For him to articulate in his own special way at the age of two his emotional need is a reminder that there is so much power in communication.

Recognising the ways in which babies communicate from birth helps us to understand their needs better and may reduce stress for the new parent.[5] Knowing that most of their needs in the early weeks and months are linked to a need to be picked up to meet them supports the concept of the "fourth trimester", which encourages plenty of holding in the first 3 months of the baby's life outside the womb. Being attentive to the developing language of the individual baby will help with identifying when they are asking to be picked up. This has a wonderful side effect of creating a 2-way dialogue between baby and caregiver.

Communication from caregiver

From our side of things, our voice can signal to a baby that we are going to pick them up. In the Anticipatory Adjustments to Being Picked Up in Infancy study they found that from around 3 months of age not many of the babies were making postural adjustments at this stage, whereas at 2 months old 50% were. It would be useful if we had more information about where they were at developmentally when the anticipatory responses stopped. My observations of this, and many guidelines for reaching developmental milestones, has shown that babies of around 3 months of age tend to have reached a stage where they are awake more and taking more of an interest in the world around them. Their "world" is opening up beyond that of their caregiver/s. Their sight has improved to distinguish between similar shades, they can see further than in the earlier days and weeks, binocular vision control is advancing and their awareness of themselves and what is going on around them is increasing. Communication from their caregiver/s tends to change somewhat when they're spending more and more time awake. The emergence of smiling and other interactions, as well their growing need of some sort of entertainment during the active-alert state means the two-way communication is developing rapidly.

I believe that the changes in communication coupled with this increased awareness has an impact on whether they will expect to be picked up or not. Babies spend much of their time eating, sleeping and eliminating in the first 2-3 months and their awake-time interactions may be more limited in the first month or so. "Chat" during this time is likely mainly happening when they need something (so would be more likely to be associated with being picked up), and when they're being held, due to the limitations they have in interacting with the world at this time. When you are more aware of what is going on around you, you're going to start associating things with an awareness rather than an inbuilt, reflexive response. If your caregiver doesn't communicate in specific ways to you that they are going to pick you up, then how are you meant to learn to anticipate it happening by being spoken to? Specific, familiar words are needed to develop an association between chat and being picked up.

So, I wonder if the results would have been different if the caregivers had been instructed to use a specific phrase such as "I'm going to pick you up now" every time they picked their baby up *before* the 2/3-month mark. Would an association develop over time as the gradual process of "waking up" to the world around them happened over a number of weeks, leading to them easily being able to differentiate that phrase from other direct chat? It seems plausible, especially as we know they understand words and phrases long before they learn how to say words as we do. We know that this definitely happens when they're older and pre-verbal even if they're not spoken to most of the time before being picked up. At some point they learn to associate certain words with actions. It's also believed that babies recognise intonation[6] which lends support to the idea that they may be able to associate phrases with actions.

It's interesting to think about how caregivers can encourage their baby's communication by modifying their own; both modelling examples of how to communicate to us and giving them time to adjust to prepare for being picked up. In a fast-paced society it's very easy to slip into just picking them up with no warning, no communication about what's going to happen. We can involve our children from the beginning of the carrying process, right from birth.

Approach

Approach is next on the list - how does this work? In the earlier months it may be that approach, in general, signals that carrying is going to take place, as when your life consists mainly of eating, sleeping and eliminating, most of the times you are approached is going to result in being picked up. Like with the "chat" phase, as you develop further, can see better and have longer periods of "awake" time, it's likely to get easier to distinguish between approach behaviours and you'll eventually work out which ones apply to carrying and which apply to, say, peekaboo.

On approach the baby may lock eyes with their caregiver, and when they're a bit older (this was noticed more at 4 months in the study) they may avert their gaze to the caregiver's outstretched hands. It appears that looking into each other's eyes helps the baby's awareness of what is happening by them being present to the caregiver's movements, observing them moving closer to them.

A quietening of the babies' movements was seen, with their arms away from their body which may indicate a response of making room for the caregiver's hands to begin the initial contact phase. Some babies also arch their back and tense their body on approach (stabilising the body ready for displacement), which is something we tend to see more in either the initial contact or lifting phases, as a direct response to the specific touch associated with being picked up or the movement generated from lifting.

As they develop further they may begin to respond to a caregiver's outstretched arms by reciprocating the action, thought to begin at around 6-7 months of age, which may coincide with reaching the developmental milestone of independent sitting. Also, we begin to see self-assertion coming through stronger and they eventually learn to resist the anticipated picking up if they do not want it. They may protest in ways such as vocalising or moving themselves away from the approaching caregiver at this point of the process of being picked up.

From the baby's point of view, once they have figured out how to move themselves across a floor in some way (e.g. crawling) they now have an additional way of communicating their need to be held. They can physically approach their caregiver. This approach is usually coupled with some sort of vocalisation and/or an attempt to communicate physically. They may place a hand on the caregiver's leg or foot, pull their clothing, or attempt to pull themselves up onto them. This new way of initiating the carrying process will continue to develop over the coming months and years as they learn to cruise and walk, and their language improves. In fact, approach often replaces other communication from them as their starting point for initiating being picked up.

Initial contact

Initial physical contact with the baby's body can also trigger anticipatory responses and I believe this is based on the *type* of touch, which tends to be a grasping action from the caregiver around the upper ribs just under the armpits. You wouldn't expect, for example, stroking a baby's face to send a signal to them that they are about to be picked up. Responses may include tensing of the muscles and reduced movements (slowing down or stopping wriggling/thrashing around). It was shown in the study from reference 1 that between 2 and 4 months the babies' coordination and smoothness of movements increased, implying that – just like in carrying – the repetition of these movements combined with a readiness for being able to learn the voluntary action enables the development of controlled, learned behaviour.

At later stages of development, just like in the previous phase, we may see a protest or disengagement from the caregiver if the baby doesn't want to be picked up. This may involve things such as vocalising, stiffening of the body, flailing or going floppy. If happy to participate, older babies may engage their muscles more at this point, readying themselves for the next step – picking up.

Lifting

Lifting is the first of the vestibular responses in picking up. The motion, along with the change in trajectory sets off reflexes of postural adjustment in younger babies. For example, in a newborn or young baby, a head which is left to hang backwards will trigger an extension of the body (tonic labyrinthine reflex), which is obviously not useful as a precursor to being carried. If the head is supported at the midline it will keep the limbs in a certain degree of flexion. This specific reflex reminds us to support the newborn's head/neck as well as their ribcage when lifting up. It also shows us what position the head and neck should be in to promote ergonomic carrying.

A baby with good head and neck control will be capable of developing the tension needed to counter being lifted up. Of course, mindfulness is still needed. The faster and sharper the movement the less time they have to anticipate and adjust and finding a balance which allows them to exercise this postural reaction with control is ideal. With the stabilisation of the head and neck comes tension in the shoulders and upper back. This benefits the caregiver as well as the baby, as it means the part of their body they're holding onto and lifting is engaged rather than floppy. A natural progression from this is an engagement of the core muscles as the legs lift up into the reflexive spread-squat position. If they are placed on their caregiver's body in this position regularly they are much more likely to associate adopting this position with carrying and learn to do this as a voluntary action as the reflex begins to integrate. If they are usually carried passively but held in a sling or carrier in a spread-squat position they are more likely to retain the instinctive adoption of this position, if not on the way up then at the next stage. This can also make it easier to teach them to actively cling at later stages of development.

As babies reach other physical milestones, such as sitting, the anticipatory responses may change in some ways. Being picked up from laying down requires different sorts of responses to being lifted from an upright position. A seated baby will either let their legs dangle passively upon lift-off, or counter gravity and adopt a squat position. They are more likely to be disrupted at the initial contact and lifting stages if they are engrossed in an activity or not looking at their caregiver, as their line of sight is now lower. A standing child is also more likely to actively bring their legs up into the position they expect to adopt on the caregiver's body if they are participating yet can again be disrupted by the picking up process.

Older ones may develop a preference in how they're lifted up. For example, from just over a year old until around 2.5 years old, my youngest would love to grab my forefingers and have me launch him up that way. This obviously required extra involvement in him to be picked up, as he would need to cling tightly with his hands. This, combined with him swinging his legs up into a spread squat position, meant he had a lot of control in the anticipatory process. No caregiver control of the upper body meant he could choose exactly where he wanted to cling to and having to use more of his body to come up meant he was doing much more work to get onto my body, therefore engaging more muscles. On reflection I've come to wonder if this may have contributed to him firmly establishing a side preference, which is something we'll revisit later.

Bodily approach

Bodily approach is very much tied into the lifting stage, but we separate them into two as there's some difference between the initial lift and the actual approaching of the body. It may be that there's minimal time between the start and end of the lift, or it may be clearly visible where one ends and the other starts. It tends to be most noticeable when a baby is lifted up from laying down as there is usually a distinctive lift and change of orientation (to upright), and the approach starting closer to the point of being upright. As the baby is moved toward the caregiver's body the spread-squat reflex tends to stay in place, or this may be the point where the voluntary action is started, making for minimal adjustments needed once bodily contact is made. In older babies we are looking for them to actively adopt and stay engaged in this position. Learning this movement as a voluntary action is crucial for a smooth transition from being picked up to sitting on the body, and to trigger clinging from the moment they engage onto the caregiver's body. If a baby or child is used to clinging to the caregiver they will most likely adopt this position almost reflexively, as it is so ingrained that they are about to participate in carrying.

There are, of course, exceptions for this. If, for example, you bring them up to your shoulder, an experienced clinger may adopt the position on "lift-off" then adjust when they realise the part of the body they are going to. Or, when being picked up by a caregiver who primarily uses passive carrying, we may not see any leg adjustment. Again, we may also see voluntarily disengagement at this stage – either seeing the first response at this point or a continuation of disengagement from earlier stages. It's usually very clear to the caregiver at this point that they will not be engaging once bodily contact is reached. It's not an initial protestation at being interrupted, and the actions of the body may be growing stronger now.

Bodily contact

Lastly, bodily contact is made between baby and caregiver. With newborns and young babies, where they are placed and how much contact is made will then trigger the on-body carrying reflexes so that they are in a suitable position and engaged in the carry to the extent needed by the amount of support being provided. With newborns, we tend to bring them up into a shoulder hug position, or maybe chest-to-chest, and with older babies and toddlers it may regularly be the hip. Babies and children tend to learn to create the anticipatory response to the position they are used to being carried in most by the specific caregiver.

The immediate response to bodily contact ends the picking up stage and begins the carrying process. Everything which happened before the bodies connect has set the stage for how this particular carrying session starts, and the more prepared they are for this the more fluid it plays out. If they're ready for carrying to begin we see a smooth transition but if they've not had much warning we're more likely to see disjointed and uncoordinated behaviours, both before and on contact.

When bodily contact is made we can not only expect to see a variety of reactions based on where on the body they are put, but also due to whether the caregiver is supporting active or passive carrying and whether the position they are put in is what the baby/child was expecting. The most contact responses tend to be seen when the baby has freedom of movement of their extremities. The extent of these will vary based on things like the specific points of contact, how many primitive reflexes are still present and how many voluntary actions related to carrying they have learned. Children used to passive carrying in a non-ergonomic position tend to display the least reaction on bodily contact.

If the child does not want to be carried, then we may see attempts to create distance between their body and the caregiver's. Actions such as pushing them with their hands, arching their back, pushing/kicking of the caregiver's body with the feet and loud vocalisation are common responses. This may signal the end of an attempt to carry them, especially if their actions are making it impossible for the caregiver to hold them safely. It can be frustrating when this happens. However, this experience can provide an insight into the needs of the baby at that moment.

Summary

So, all these stages rely on communication of some sort, whether that be vocally from baby and/or caregiver or from the stimulation of various sensory systems. We've also ascertained that anticipatory responses seem to be reflexive in the first instance and are learned behaviours as they develop further. The responses cultivated as muscular and motor development show they may vary based on their expected outcome from being picked up.

There is also anecdotal evidence that babies and children learn to make different anticipatory actions based on *who* is picking them up, in relation to how that person usually carries them. We can observe this in everyday situations, just watching how they respond to different people picking them up. It's particularly noticeable if they're used to active carrying. This is interesting, as it seems reasonable to expect a baby used to clinging to automatically cling to whomever picked them up. So why does this happen? A plausible explanation may be that a baby is effectively conditioned into their responses based on their early exposure to carrying. For example, if they have two people who carry them regularly and one carries using active principles, they will come to learn to cling to this caregiver. They will not learn to cling to the second person as they aren't given any opportunity to.

Something to note regarding the research to do with being picked up is that it was conducted on mother-baby dyads. Whether there are noticeable differences when a father, other caregiver or someone less familiar goes to pick them up is something worth looking into in the future. I think we would expect to see similar adjustments in reaction to regular caregivers. I speculate that there may be a significant difference in response to stages before bodily contact is made from unfamiliar people. I would also expect to see a higher incidence of negative reactions from more self-aware babies, exerting their ability to communicate that they're uncomfortable with strangers or people they don't know well entering their personal space and touching them.

We may also see different postural adjustments when being picked up by different caregivers who regularly carry them. The way in which babies are picked up is going to influence the type of anticipatory response. For example, one person does a straightforward lift, and another does a novel "flying" lift. In the "flying" example, we would at least expect to see a postural response of the landau reflex being triggered in babies who have developed it, up until integration. If it becomes a learned behaviour we're likely to see the child continue these adjustments as voluntary actions going forwards. We would expect to see responses related to an upright lift in the generic way of being picked up.

So, it's obvious that participation in carrying can begin with an expectation that they're about to be picked up. However, we also know that bodies also make reactive adjustments when they are destabilised. Providing the time for anticipation to occur makes the process of picking up smoother for baby and caregiver and causes less disruption to the baby. It also reduces discomfort and disorientation. How aware caregivers are of this process, coupled with how they carry them when they're on-body, lays a foundation for their learned responses going forward. It's possible to nurture this by adopting practices which allow for anticipatory responses. A baby who is actively involved in carrying before being placed on the caregiver's body is even more aware of the process and is better adapted to adopt a position suited to clinging. The responses also mean they're engaging their muscles as they are lifted which helps take some of the strain off the caregiver during this time.

Caregivers are able to develop their own anticipatory routine based on their parenting style and the baby's preferences. They may decide to use an approach involving verbal communication to give their child warning that their play is about to be interrupted, for example. Or, like Isaac and I, they may choose a novel way of making lift-off more enjoyable. An awareness of the anticipatory process and potential responses may also aid caregivers in recognising how the baby's development in carrying is progressing. For example, noticing a delay before the legs are raised into a squat position may indicate that the reflex is being integrated and the voluntary action is being learned. Or, if a baby isn't adopting this position at all the caregiver may choose to verbalise what they want them to do, whilst physically adjusting their body into position.

Interestingly, parents with children who have autism have reported that their child/ren do not make anticipatory responses to being picked up.[7] In my first book I touched on the fact that people on the spectrum have been found to have retained or missing primitive reflexes which in one study a suggestion was put forward that the diagnosis – which tends to happen later in childhood – could potentially be given much earlier if specialists are looking for other things such as reflexes.[8] I'm repeating this here as it's an avenue which has already been looked into by researchers, so it would be interesting to see what may be uncovered if a similar study were to be initiated, looking into potential markers flagged up by the carrying process.

Chapter 2.2

Developmental Process

Active carrying/clinging is a developmental process!

I cannot stress this enough and look forward to the day it is finally recognised as such. It is a part of normal physical development that is either partially completed or missed out on in so many children, and some of the questions this raises are: "What does this mean physically for the ones who don't complete this process?", "What does it mean for those who do complete the process?" and "Why is it so important they do so?".

At this point in time the answers to these questions are very limited, as no specific research has been conducted on this subject, clinging in carrying is not recognised as something normal and, of course, active carrying/clinging is not yet recognised as a developmental process. Obvious answers include:

- Not completing the process leads to less than normal bodily strength and clinging-specific movement of the whole body (and by normal, I mean normal in terms of strength and tone when the clinging process *is* completed)

- Those who complete the process achieve normal bodily strength and movement *on their unique spectrum of clinging capacity*

- It's a normal developmental process, not an added bonus – we recognise the importance of achieving all other developmental processes so why should the carrying process be any different?

We obviously need much more research into this than observations and seemingly obvious answers, to dig deeper into this and find out what normal clinging behaviours look like in our own cultures. What is normal for one culture may not be for another. There may also be subtle (or not so) differences in how babies are supported and/or the way they behave on-body, linked to certain cultural practices, for example. We may not be able to give definitive answers to these questions as yet, but we can look at this process as a whole to gain deeper understanding of how it works. In this chapter we will look at *why* active carrying/clinging is a developmental process and the stages it goes through from reflexive to independent clinging.

Motor development

Human motor development is a lifelong process which evolves over various life stages. It's also something which is dependent on other developmental processes – the intertwining nature of general maturation is undeniable. For now, we will focus primarily on motor development – how it presents in carrying and how other areas of physical development impact clinging development. In later chapters we'll look to other areas and see how they are affected by or impact on carrying.

Just like all other developmental processes, clinging requires countless repetitive actions to be made over weeks and months in order for the baby to learn and become extremely capable at replicating the movement. Just as babies learn every other aspect of physical development this way, we cannot expect them to know how to cling if they're not being exposed to opportunities to do so. As clinging is not limited to on-body, they have other opportunities to practice similar actions off-body which may aid in the carrying process, such as grasping, climbing and squatting.

Although babies are born preadapted to cling, they must first explore and get used to the landscape of the caregiver's body, to familiarise themselves with their carrying environment and work out on a certain level how they fit to the carrying person. As clinging behaviours develop, physical development increases, positions on the body change, and the physical makeup of the caregiver may change over time. Babies have to adapt to the new environments presented to them, and gradual transitions seem to be written into the very nature of clinging. As we saw in chapter 1.2, clinging naturally evolves over time as the body grows, voluntary actions are achieved, and strength develops.

Similarly, babies learn to interact with and navigate the world via static positioning on the floor or in devices, then develop ways of moving further through rolling, creeping and crawling. It all happens gradually, and yet when they can pull to standing and cruise, then begin to walk, the nature of interaction with the world goes through yet another change. They're able to engage in new ways and from a different perspective. This also happens in carrying when they transition to the hip, then as they have the bodily strength to sit in a shoulder carry, and again when they are able to cling to the caregiver's back. New situations arise, yet each previous stage has prepared them in some way for it.

If we're to encourage a smoother process, ensuring reflexive to voluntary clinging happens seamlessly, then we are responsible for providing the right environment for this learning to occur. Just as we put the pre-crawling baby on the floor to practice the many movements involved in what will eventually evolve into crawling, we likewise support the baby on-body in active holds to practice the reflexive actions which will gradually transform into voluntary, stronger, more fluid and varied clinging behaviours. We don't expect them to go backwards in any other area of development, and so the same should be for carrying.

Human beings learn through repetition, whether adult or child. We know that "normal" child development in this day and age would actually be considered delayed development decades ago, as the time it takes the "average" baby to reach milestones has grown so much.[1] It's been put forward that this is down to the increasing amounts of time babies are confined for, whether it be in a car seat, bouncer, stroller, baby carrier etc. Even playpens – think about the distance babies cover when they are free to roam, and how confined spaces limit that. We know from observations of animals placed in captivity that restricting the natural environment affects their development.[2] Whilst there are many obvious differences between humans, non-human primates and other animals, we know that varied movement is required for normal human growth, so acknowledging the detrimental impact of restricting the environment of other creatures isn't venturing too wide of the mark.

In a similar vein, the restrictiveness of using only passive holds stunts normal on-body clinging development by removing the opportunities to interact with this natural habitat in the way human young were designed to. We were *born to cling* to our caregivers. It's as simple as that. Reflexive clinging is not something of a fluke which is meant to only provide a few months of help. It's merely the extra help we've been given to make carrying possible before babies have sufficiently developed to have better control of their body.

The carrying process begins right after the bodily contact which prepared the baby for the movement of carrying. In fact, as we touched on in the last chapter, the beginning of the carrying process merges with the end of the preparatory process. We can break it down further into subcategories of "passive", "partially active" and "active". I tend to usually just refer to passive and active rather than include the middle-ground, purely to keep things simpler, especially when "active" and "passive" have different meanings based on the type of position/hold they are in.

As we already explored, this contact fires up the reflexes or voluntary actions (depending on their stage of development) for the type of carrying being initiated. Let's begin by looking at the basic definitions of "active" and "passive" carrying, and then we will explore what this actually means for different holds during the developmental process, and I will provide information on the relevant support progressions to aid their development at different stages.

Passive carrying

This type of carrying involves the caregiver providing most or full support for the baby/child. It's usually defined by full-body contact, with the baby/child in a relaxed, non-clinging state. The caregiver holds the "load" in close contact, thereby minimising feelings of additional weight (the further away from the body, the heavier they feel). Full-body support is provided by providing a "seat" for the bum and (but not always) legs, as well as either physically supporting the upper body or using our body as a resting point.

Reflexes and voluntary actions are still triggered in this type of carrying, though not as much, as the baby still aids the caregiver by positional adjustments. The very nature of the support given, however, signals to their body to relax. Adjustments are mainly made by the caregiver and the more we use passive carrying as a primary source of carrying, the more we program our children to be passive. Passive carrying is holding. Holding is very much suited to when they are tired, asleep, or need to disengage from the world, and using it at these times means we're using passive carrying in the way it was designed for.

Partially active

In this way of carrying we see full support of one half of the baby/child's body. This could be just providing a "seat" for the bum/legs so that the upper body is active or pulling the upper body to the carrying person's and leaving their legs unsupported, so they cling with their lower body. In front-clinging this is characterised by cuddling their torso with one or both arms, providing full support of it, whilst their legs actively cling. In the shoulder hug it's supporting the bum and legs whilst the whole upper body is unsupported. We refer to this as "active carrying" in this hold though, as it's not possible for them to cling in this position due to no anchoring point on at least the side nearest the midline of the body, but usually both.

Active carrying

Active carrying is enabled by either providing one point of support or none. If providing support, this is done in a way in which simply provides a point of contact and doesn't bear a great amount of weight. It tends to be an arm at some point on their back and does not entail pulling them tight to your body. It means that they must actively cling with their lower body, work their upper body and usually create an "anchoring" point with a hand or arm. It's defined by much lower body contact with the caregiver, but lower legs and feet may differ in contact during carrying for a variety of reasons, which we will explore later in the book. In hip carrying, the upper body tends to have much less contact and usually turns away from the caregiver's body to a noticeable extent.

> *If providing no support, we see an exception to the "rules". Most independent clinging is characterised by full-body contact – the extremities are used to hold onto the body and all the work is done by the baby/child. This requires the full contact to make it easier for them to cling fully, and as an added bonus it makes them feel lighter to the caregiver. The caregiver provides no support whatsoever and is merely an "object" on which to cling.*

Newborn to 12 weeks

As we know, the carrying process begins reflexively. For a term baby with no developmental or medical issues which may affect carrying, we can expect to see the carrying reflexes in full swing from birth. On-body carrying behaviours can be seen from the moment baby is brought to the caregiver's chest or stomach once they are born, and lots of attention has been given to the "breast crawl", which we looked at in chapter 1.3.

From my own memory growing up, it seemed to be that the "right" way to hold a newborn baby was to cradle them in your arms so that they were laying down. This is how I began holding my firstborn but quickly realised she needed to be upright most of the time otherwise she would cry. In the 14 years since, I've rarely come across a baby who doesn't prefer to be held upright. Although some carrying reflexes are triggered in a cradle hold, we see many more in a shoulder hug or chest-to-chest. I believe this is to do with things such as being on the vertical plane, the amount of contact with the body, that they are less passive positions than cradle, how the baby fits to the caregiver's body, and the effects of gravity on their body.

When newborns want eye contact they tend to be held in their caregiver's arms away from their body, face-to-face, rather than in a shoulder hug or chest-to-chest. A separation of the torsos occurs and for this to be possible without unrest from the baby they must be relaxed and content, which is most often in the quiet alert state. Active carrying of brand-newborns tends to happen in both the quiet-alert and active-alert states as the activeness is primarily of the head and neck. As the newborn develops head and neck control this tends to be more in the active-alert state, as they are still happy enough but indicate they are in need of stimulation. Quiet-alert is inherently still, with little movement, and a good time for eye contact. It's better suited for off-body interaction or relaxed carrying.

The active-alert state of alertness tends to be combined with fussing when described by others[3] which is why I refer to 7 stages rather than the generally accepted 6. Fussing is a precursor to crying and indicates an unmet need, whereas being active and alert indicates a need or readiness for interaction and movement. It's important to separate these differing states as the definition also has implications on the perceived abilities of the baby to interact and learn. For example, although the scale is aimed at newborn behaviour, babies, children, and even adults spend time in differing states throughout the day – the duration just changes. As babies get older they stay in the awake states for longer periods and it's easier to differentiate between them. In the active-alert state the baby seeks engagement, their primitive reflexes are readily elicited, and the infant is primed for learning from the world around them. If we add in the fussing state, this suddenly changes to the inclusion of erratic movement and disengagement from the environment, which is not conducive to participation in carrying. As you can appreciate, there are marked differences between being active and alert or fussing.

Although the term "newborn" is defined as birth to 28 days of age[4], we will define the "newborn" period as between roughly term and 6 weeks of age, as many caregivers lean towards this definition, as well as differences in the carrying process emerging from around this time. As stated earlier, from birth the baby's reflexive actions are a continuation of movement from life in-utero. By 6 weeks of age, the term baby is usually still strongly dominated by the flexor position of all extremities, but in over the next 6 weeks or so we see a clear emergence of extensor tone in the neck and upper extremities.

Carrying vertically seems to be biologically normal from birth onwards, and we know that this is safe for their spine.[5] We also know that head and neck control are practiced in an upright position. As babies are born needing to build up their muscular strength and complete various stages of spinal development, they are relatively floppy and need help and support to stay in various positions as they develop strength and control. Head and neck control vary between babies – even from birth – and is a good reminder that everybody is unique and develops at different rates. Biological maturity cannot be linked solely to age. We also all have different spectrums of physical ability. How we support a newborn baby in chest-to-chest and shoulder hug holds will vary based on their head and neck control, as well as their state of alertness. Let's start with the passive positions and work our way up to active.

Shoulder hug

This tends to be a go-to hold for many caregivers and it makes sense because our bodies are conveniently designed to support babies off-centre. One arm either side of the body provides a support mechanism on each side.

In a shoulder hug we can support their bottom and thighs with the forearm that corresponds to the side of our body we have placed them to. Keeping their bottom central on the forearm will help protect wrists from strain injuries. The other arm/hand can then support the baby at an appropriate place on their back, which we will come to in a moment. The way our arms are designed means that when we hold it to our side and bend our forearm up at the elbow we cover a nice distance across our bodies if we angle it diagonally. In fact, it's the perfect distance to use a hand to support a baby's back or head/neck! Off-centre, it is much easier to seat them centrally on our forearm, or at least away from our wrist. This can help minimise shoulder strain, especially when passively carrying older children. Shoulder strain is more likely in passive chest-to-chest carrying with older babies and children due to how the arm must move round and align itself with their wider bottom.

In active carrying, a newborn/young baby will be a little lower on the body so there is still great body contact, but their body is in a straighter position, allowing them to lift their head up and strengthen their neck muscles. At this stage we would tend to support them around the upper back with our free hand. As they gain full head and neck control, that support lowers to the mid-back to allow more freedom of movement for the upper torso, which is now the focus of strengthening. When full torso control is achieved we're able to support this hold in the most active way possible – by just using our forearm for them to sit on, meaning they are fully active in their upper body. At 3 months and younger, babies tend to enjoy this position in both the quiet-alert and active-alert states. In quiet-alert, they're able to quietly observe the world around them at a good vantage point. They can take in their surroundings whilst connected to their caregiver through touch. During the active-alert state they are likely to move more and engage with the environment; for example, "talking" to inanimate objects they see. In this state we see carrying more as exercise and movement than in the quiet-alert state.

Depending on whether the baby is in an active-alert state, quiet-alert or is tired or asleep, we can decide how much support is appropriate. For a baby with very little head/neck control in the quiet-alert and active-alert state we will be wanting to provide a point for them to rest their head on, but also giving them enough freedom of movement so they're able to work those muscles. This tends to work well by holding them with their hands on the carrying person's shoulder, at a height where the head would rest closer to the front of the body than fully on the shoulder. This provides a "cushion" of sorts for them to lean their head against but isn't angled to a point where it discourages them supporting their head. Many repetitions of these movements over the days and weeks ahead will contribute greatly to the baby gaining full head and neck control, at which point they will be ready to transition to the support appropriate for encouraging upper torso control. At this point, moving the support of our hand to the mid-back provides the freedom and encourages the movement needed to work the upper torso. Again, the point of support is right next to the unsupported part of the body so it's easy to quickly adjust back to the previous level of support if they become unstable or tired.

Chest-to-chest

This can be a fairly instinctive position for caregivers, especially when sat down or reclined. With in-arms carrying it appears to be more popular with people who also babywear. This is likely because most newborn babywearing in Western cultures is done chest-to-chest, so there may be a subconscious direction to let babywearing position dictate in-arms position. In carrying, this position seems to be better suited for reclined holding and practicing skin-to-skin. This is because the proximity of milk to the baby is very close in this position, so in general carrying it can be counter-productive, as it can be a distraction. Their viewpoint is limited to left, right and being supported to look at their caregiver's face, which provides limited interaction with the world. Also, the shape of the body on which they are supported can encourage a more relaxed position if breasts are of a certain size. This curvature provides more of a resting point than a flat body. With these points in mind, chest-to-chest may be more suited to the quiet-alert state in the younger baby.

Another limitation is where the arm position falls when held at a right angle. The first half of the forearm rests in the middle of the torso, and this is the worst place to bear weight as it puts strain on the wrist. To be able to make use of a forearm closer to the elbow the arm must come further inwards, which creates strain on the shoulder joints. Arms can also be combined for support, of course, but when they're younger it works best to support their legs and bottom with one arm and their back with the other until they're big enough for the caregiver to use each arm to support each thigh. At this point they will have enough torso strength to support the rest of their body themselves.

Once upper torso control is achieved

Flexor-dominant tone continues at the 3-month mark, but extensor tone is rapidly developing, especially in the neck and arms. Once upper torso control is achieved, babies are able to spend more time in the prone position, holding their chest and head up by extending their arms on the floor, which aids extensor tone. Independent sitting at this stage is not possible due to the lack of extensor tone in the trunk and hips. Active carrying facilitates development to independent sitting due to the opportunities to develop the core muscles, as well as the spinal muscles responsible for this endurance-related posture. At around this time the Landau reflex emerges[6] which increases extensor tone, especially in the lumbar spine. This isn't a primitive reflex as it emerges post-birth. Its significance in both carrying and physical development in general is that it contributes towards the development of the lumbar curve in the infant's spine. As we know, spinal development is sequential from the head down. Having a reflex which curves the lower spine outwards from birth would be counterproductive as the relevant upper body strength and development isn't there to support the upper torso.

It tends to be elicited off-body as it's triggered in the prone position and is regularly in action when the baby is playing on their front with arms extended to hold themselves up. Although this isn't a carrying reflex, it directly impacts on development related to carrying. It facilitates longer periods of upright posture by activating the deep postural muscles of the neck and back; development of which enables progression to achieving independent sitting. Carrying aids the lumbar spinal development by enabling the baby to sit in an a straighter position on-body than on the ground, whilst still being supported appropriately. The difference between using a device to prematurely sit babies up and enabling it through carrying is that on-body the baby is engaging in the process of "sitting" on the body. The caregiver also provides intuitive and appropriate support, which can also be varied based on the baby's needs, and weightbearing occurs in a different manner when in off-body devices.

Transitioning to hip

The time to transition to the hip will depend on both the baby's stage of development and the comfort levels of the caregiver. To carry actively on the hip the baby needs very good upper torso control, due to the disconnect of the upper body. The mid- to lower-back requires support before they can sit unaided to make the process more comfortable and developmentally appropriate. Although they are now well-versed in being active with their upper body in carrying they are now experiencing *clinging* for the first time. The dynamics of carrying have changed and physical development is occurring alongside this. While it is technically possible to support a baby at earlier stages of development higher up on the body, it means carrying will be less coordinated and the lower extremities won't be clinging as well (as the lumbar spine is still underdeveloped) and the caregiver will need to pull the baby in close to their body to compensate for the unstableness. This overworks the lower body as it has not reached the stage in which it is designed to comfortably sustain this sort of exercise for extended periods, and there is great potential of compromising the upper body/spine by using the force of pulling to bring their upper body inwards whilst also bearing weight through this action.

Other things to look out for which indicate readiness include the strong presence of the postural labyrinthine righting reflex when the body is tilted to the side and the separation of whole-body reflexive actions into upper- and lower-body reactions. For example, a clear Landau reflex presents with an extension of the upper extremities and a slight flexion of the lower. Top and bottom are able to produce opposing actions rather than identical. The ability to keep the head in line with the body enables good posture on the move and when, for instance, the carrying person leans in a certain direction.

The baby also needs to be able to use their arm closest to your back to "hold on". This means they need some amount of control over the arm or assistive reflexive movements. The ATNR may or may not be present at this time, but if it is it should be in the process of integrating. As we saw in chapter 1.2, the upper extremities reaction lasts longer than the lower. Babies tend to mainly look away from the caregiver's body than trying to see behind it, so the most likely action to be triggered is a hugging of the carrying persons arm which assists carrying. The arm not in use may be engaged in pointing and reaching but is also likely to spend periods of time relaxed.

To gain this arm and hand control babies practice reaching out for objects and people, and this results in smoother and more controlled movement. These motor skills are usually developed by around 5 months of age.[7] This links closely to the integration of the ATNR as independence from this reflex enables voluntary positioning for hand function. Voluntary neck righting is also present - being able to move their head without a reflexive whole-body "righting" response enables interaction with the outside world from the hip without their body trying to follow their head.

To reiterate, at this stage, by providing support to the mid- or lower-back (depending on individual strength and development) they are able to balance out their overall strength and clinging capacity in a way that means the point of support the caregiver is providing is just that – a point of support. They *never* want to be in a position where they are having to press against the baby's spine in a pulling and weight-bearing action. The signs of this are quite obvious – a tightening of the muscles in the arms and a strain on the bicep, feeling the hardness of their spine against the forearm, as well as the aforementioned awareness of having to pull them in and feeling a change to weight-bearing. A need for this sort of support in hip carrying is likely to indicate the baby is not ready for this position or is too tired to participate at a high enough level.

It's interesting how by the time they are ready to transition to the hip they tend to be at a weight where creating a seat with the arm is likely to be getting more cumbersome, and the position has developed into one more suited for when they're tired or asleep. On average, breastfed babies tend to reach double their birthweight at around 4 months of age.[8] This is coinciding with good upper-torso control and a developmental point of working on the lower torso strength to gain full torso control.

There comes a point in the shoulder hug where the passive style of carrying may become more comfortable as – although the upper body is working in the active hold – the arm is having to bear all the rest of the weight. It's the legs which do the bulk of the work in clinging, and this isn't possible in the shoulder hug position. In the passive hold the caregiver's body distributes more of the weight-bearing to the back and shoulder muscles by the way they position the baby higher up and over the shoulder - they take more of their weight due to the angle the baby is placed at. This means the forearm is required to do much less work.

Remember, passive holds discourage activation of reflexes – when babies are active and alert, the aim is for active participation in the carrying process, and by regularly using passive positions during this state (which is highly conducive to learning) we can effectively condition them into passivity. Moving to the side of the carrying person's body is a stage in the developmental process where babies now begin to learn how to use their stronger body in different ways. This is a whole new way of behaving on-body, and being supported, and it's something they will obviously need to get used to. The carrying reflexes still present at this time make the transition easier for them, but it may feel a little strange to caregiver and baby at first. Their body still has lots of work ahead of it, including more spinal and muscular development (especially of the lower torso and extremities), the integration of clinging reflexes and the emergence of their voluntary actions. This means that clinging will strengthen over the months and years ahead, which is needed to enable the baby/child to continue to cling as it gets heavier. Endurance is built by both baby and caregiver by practicing active carrying.

Independent sitting

I found it very interesting when I researched current information for encouraging independent sitting. Parenting websites distribute advice such as "tummy time" before the baby is able to hold their head up, putting them in Bumbo seats from 3-4 months of age, propping them up with cushions in a seated position on the couch and so forth. This is – apparently – to get them used to these positions and encourage development of muscles necessary to unassisted sitting. We've looked at how readiness is needed to be able to attempt achieving a skill and how passive positioning in containers isn't useful for muscular development.

To me, from my unofficial observations, it's always been clear over my years as a parent that strong neck control tends to come easily for the majority of babies, even when the caregiver/s have rejected "tummy time". In these babies there's been a strong correlation with upright holding positions and/or babywearing and good muscle tone. Upright positions mean the baby isn't having to work against the forces of gravity on their head anywhere near as much as when they're horizontal. This enables them to focus on endurance and control rather than bursts of strength to get their head off the floor, for example. In a similar manner, being carried in-arms provides sequential developmental support for muscles used in sitting unaided, right from birth. Sitting practice therefore can come as a by-product of active carrying.

Babies are designed to be carried in-arms and – as we will see with many other designs of the body – nature is very efficient with the number of uses and benefits for any given part of the body or normal, everyday functions, movements and practices. From an evolutionary point of view, it makes no sense that caregivers would need to carve out extra time in the day or use man-made devices to ensure babies' development. Yes, we absolutely can add to their experiences by choosing to use certain products, but we must recognise that (for most humans) we are designed to thrive by interaction with caregivers and varied environments. By viewing the caregiver's body as both the first environment a baby interacts with and explores, and its continued major role once they're ready to discover more of the world, we can see better how it is designed to aid their development. Babies are held in upright, seated (or squat) positions from birth. The way active carrying transitions aids muscular development, engaging the muscles of the body (including the core) to help reach the next stage of physical development.

It's in the nature of sitting for the spine to be straightened. Normal off-body sitting happens via sitting on the ischial tuberosities (sit bones), which provides a base for aligned sitting. This takes pressure away from the lower spine. Though sitting is a position of rest, it still requires the person to be fairly active in their upper body if there's no support for the back. Even though we provide a support against the spine in active carrying, it's also in the nature of this practice for the baby to sit straighter. The more the spine develops, the straighter they will usually look in an active carry. This links into the integration of the squat reflex, a moving backwards of their pelvis, and how their legs also begin to straighten out more the longer they've been able to sit unaided, going from knees bent outwards all the way through to sitting with legs straight out in front of them. When they sit with their knees higher than their hips it encourages curvature of the spine. This in turn encourages more contact of the torso which can trigger the baby to become more passive in carrying. It's for this reason we tend to often see a less pronounced squat in hip carrying, with the 90° angle being favoured by them around the time sitting is achieved.

At this point support can now be moved further up the back. As they tire, lower support can be tried out to see if they're still happy to continue with active carrying. If they are, this support can either be used until they tire, or the caregiver can switch up the points of support to encourage periods of greater clinging and rest. Something important to remember is that there is a dominant leg in hip carrying, so changing sides also provides relief for the leg which is doing most of the work, as well as the caregiver's body.

A partially passive chest-to-chest hold also works well at this time, as the excellent full-torso control means that both arms can be used to support the legs, either as a double-hoop support (each arm hooping some way around the bottom without straining the shoulders) or one forearm supporting each thigh. They are active with their upper body as the lower body rests. Another variation is supporting just the spine to keep the legs free to cling - it also provides an even distribution of work for the child's legs. This can work well if there's good alignment of the caregiver's shoulder and the baby displays an ability to cling to this wider surface. Oftentimes they will need to wait for their legs to grow longer before it becomes easier to do, as the nature of the front of the body is different to the hip where they "slot on" more easily. It's important to remember, though, that positioning on the front of the body like this can cause downward pressure on the pelvic organs – more so if there are postural issues – so is one to use with caution, especially if there are postnatal conditions present.

It tends to be between the sitting unaided and crawling milestones that we see the emergence of active disengagement from carrying with the legs. This is as the initial squat reflex integrates and they gain voluntary action over the position. Its significance extends to other areas of motor development as control over flexor and extensor tone is required in crawling, then standing and walking.

Crawling

When babies are able to sit well unaided, the focus transfers to strengthening the legs even more. The next major physical developmental milestone is crawling, and this requires the ability to support their weight through their arms and thighs. To make the move towards this they're required to exercise their arms and legs in various ways. They've already been building arm strength from supporting themselves on their forearms and pushing up with their hands, as well as by using their arm to support their upper body in carrying, and the next step tends to be rocking backwards and forwards in the stationary crawling position. Whole-body extension and flexion reflexes have long since integrated by now and, as we found out in chapter 1.3, a precursor to learning how to crawl is the emergence of a new postural reflex – the symmetrical tonic neck reflex (STNR). This allows for flexion of the legs whilst the arms extend when the head is lifted up when on all fours. This helps them to adopt the position needed to eventually initiate the movement which is crawling. The rocking tends to involve the bum dropping down too, so practice of lifting the bum up from the floor is going on also. This is enabling significant muscular development and links into carrying so well. At this point in time, the legs are beginning to take on a more active role in carrying – voluntary clinging is developing from reflexive. Again, the role in which active carrying and learning to crawl benefit each other is clearly apparent. Stronger legs are needed to advance both.

To facilitate this, support may be focused at the mid-back and they can be switched up with buffering by the underarms as and when the baby's strength allows it. As before, an awareness of the appropriateness of each level of support is guided by what the arm is needing to do to make clinging at that point possible. Again, there should be no weight-bearing under the armpits or pulling the baby to the carrying person's body – the arm is there just as a place for them to lean slightly on. Caregivers aren't meant to take the strain; active carrying requires the continued work of the baby.

The ability to move more easily from A to B can signal a change in the nature of carrying for the dyad. They may find that the amount of time in-arms reduces somewhat, as the need for practicing the crawling-specific movements are met and the baby's world is opened up further. It's all part of the journey towards achieving full independence of movement. Although carrying is movement, it's only one part of the bigger picture of physical development. Each aspect of it is designed to work harmoniously with the others and this can be achieved fairly easily if we create the opportunities for them to move in different ways.

The newfound autonomy which comes with crawling is a huge milestone towards gaining independence. Being able to move themselves from one point to another with ease presents pros and cons though. On the one hand, they can go where they want (within reason) and don't have to rely as much on communicating to another person the need to move and the harder task of letting them know *where* to go. However, they are still very much in infancy. Their ongoing emotional needs and requirements for physical contact must still be met. This new way of exploring the world also comes with challenges, frustrations and, sometimes, injury. Carrying is still a big part of their lives and will be for a long time to come.

Cruising and walking

At this stage of development, it is crucial that the high flexor tone in the hips and knees seen prior to this point have integrated. Standing, cruising and walking cannot occur if pressure on the feet elicits flexor responses. These reflexes have integrated by way of the gradual emergence of voluntary disengagement from the flexed position which is clearly seen on-body during carrying. Active carrying continues to see a chair-like position being adopted by the baby on the hip.

As their clinging ability grows the support can be moved up to under the armpits. This tends to become a regular place of support by the time they begin to walk unaided. At this stage they may still require some amount of mixing up support points as they transition to clinging more with their legs, but many people find this eventually ends up being the preferred support point. Interestingly, as the baby grows taller the supporting point goes higher, which keeps the carrying person's arm at a comfortable height to support them at the preferred position on their body. Everything is linked in a marvellous way – growth, development and the comfort of the caregiver.

The usual occurrence of arms held out to aid balance in early walking is something babies have obviously had much practice with on-body during active carrying. From here on out we observe continued improvements in clinging as they grow taller and become even heavier. It can seem a bit confusing as to why they don't possess their full clinging abilities earlier on, especially if we try to link strength of clinging with endurance. However, they're able to cling as needed at each point in time, and for long enough periods of time. The stronger/harder clinging emerges due to the increase in their weight, meaning they must counter gravity better.

Further development

Once a toddler is walking well and the reflexes of the foot are fully integrated we may see some changes in carrying. They may want to be carried less with the ability to explore the world on foot, but it can be expected that carrying will likely occur regularly for at least the next year or so. The world is still so much bigger than them and they still have many emotional needs to be met. Growth is slowing down now – the most rapid development has occurred in their first year of life. The way clinging progresses now is still happening through carrying, but the amount of time they're spending walking and squatting is contributing heavily towards the strength of their legs.

As their torso length increases we will see a reversal of points of support over time, as the neutral arm position connects lower down the body the taller their upper body becomes. This is interesting in terms of how the occurrence of carrying slows down as they move through toddlerhood to the pre-schooler years. It almost seems like less time carrying is protected by a little more support, yet their clinging abilities are not hampered by it as their weight is still increasing and they must continue to adapt to it.

More refined and complex movements are developed as the child matures. Postural adjustments have emerged to counter for combined movements. For example, the child is now able to perform a squat whilst simultaneously playing with an object, eliciting perfect balance to execute both at once. In carrying we may see more complex actions from the child such as attempting to use both hands to play with something or complete a task on-body. To begin with this will require a change-up of support and the carrying person will notice a marked difference in their weight as they remove their stabilising arm. Over time they will be able to reach a point where they can cling more effectively with just the legs, and if clinging develops even further the older child may reach a point in time where they're able to cling to the hip independent of support, with the aid of crossed ankles around the caregiver's body.

As a tie-in to this, I should make a mention their legs in general. As they grow long enough to comfortably reach past the caregiver's outside hip, we will sometimes see them using a heel to hook over it, either as a temporary resting measure, or as a new regular way of positioning. The arms and hands of an older child may also be used to hold onto the caregiver's shoulder/s, and this is an example of how active carrying can change drastically as they get older. Before, a disconnect of the upper body was a common occurrence, but when both arms are used in either independent or partially supported clinging we see full-body contact. This sort of clinging lines up with increased use of the upper body in play. A physical ability to pull themselves up onto things with their upper body and bear more weight in vertical climbing enables them to use their arms in different ways during carrying.

As the end point of the carrying developmental process is independent clinging, we could say that it is achieved when a child can cling to the caregiver's front or back by themselves or with just their legs on the front, back or hip. The clinging process, however, will carry on for much longer as it continues to develop off-body in activities such as climbing.

Chapter 2.3

Communication

The communication present in the carrying process happens in various forms. Communication is essential to make clinging possible at all. Many of these messages are sent and received on a subconscious level, and we're going to look at this here as well as some forms being explored further in later chapters, based on *how* this works. It doesn't end there though. Active carrying promotes interaction between baby/child and caregiver outside of the need to communicate the clinging needs - it's in the very nature of the state of alertness needed to make active carrying possible.

The positions in which we actively carry on the front of our body and hips are inherently communicative. Babies are automatically positioned to be able to easily see your face which makes communicating with each other easier. Even in a chest-to-chest (C2C) position this can be facilitated as an active C2C carry requires them to be big enough and strong enough to cling well to the wider surface which is the front of the body, which tends to be once their legs are long enough to enable at least their heels to hook around the sides. At this point in time their torso is generally tall enough to automatically position their head at a height where there is less bending of the caregiver's head needed to communicate eye-to-eye.

On the hip, even when they're smaller, the disconnect of the upper body places them at a better distance to easily interact without straining ours and their neck round too far. The interaction with the *environment* also means we're unlikely to be looking at each other at all times. Clinging encourages them to use their range of vision more due to the vantage point provided by the angle created from body disconnection. In sedentary societies we don't tend to use our long-range vision as much, as we are in open spaces less often, and this is becoming a big issue.[1] For me it's been an interesting realisation that I look further into the distance and scan my eyes around so much more when I'm carrying. It just isn't an activity that you tend to do by looking down at the ground, and it dissuades you from doing things such as scrolling on a mobile phone. Carrying can create more of a presence with the world. For both clinger and carrying person, having more to see can also provide more opportunities for communication and learning.

Some types of communication we may see in carrying are:

- Reaching out to objects and/or people, including little people much further down from their height

- Attempting to engage others in conversation

- Using their body to guide the direction of the caregiver

- Babbling or talking to the carrying person

- Touching the caregivers face or other body parts

- Disengaging from the carry (e.g. cuddling in and relaxing legs or stop clinging and straighten legs)

- Attempting to find milk

- Using physical communication to signal other needs, such as needing to eliminate

- Using their body to make large movements, enabling them to signal clearly that they want to look at something

- Pointing to things of interest

- Touching things within reach

Let's explore these further.

Physical communication in carrying

The communication needed within the carrying dyad to enable clinging mainly comes from messages sent from physical contact. Many of these are sent and received subconsciously, which helps clinging to happen smoothly and without a need to be fully aware of every little thing happening. We'll cover this part of the communication in "Senses and Carrying" chapter and for now will look at the conscious and voluntary communication beyond this which accounts for behaviours such as us adjusting their position and them relaying their wants and needs to us.

Many caregiver's have a side preference in carrying. This is something I feel would be useful to look into in terms of whether or not it's *our* preferring to carry on a particular side influences preference in the baby/child. Are we communicating to them each and every day that there is one side of our body that is "right" for them to cling to? I know that I have a right-side preference and my youngest developed his preference at around 18 months old. Unfortunately, I hadn't thought to monitor something like this so have no way of knowing if there were any specific influences here, as I've always mixed up my carrying with him more than most do, with still a slight preference for the right unless I was very present to which side I was carrying him on.

I do know that it's common for caregivers to report that their baby has a preference from as young as 6 months old. This even comes from caregivers I've made aware of the implications of preference in carrying. They have noted their own preferences and then observed their babies' behaviours and communication when they start supporting on the other side for either longer periods or for the first time at all. Each person has also wondered if it's been their preference which has impacted on the baby's preferred side. As hip carrying is not symmetrical – the leg on the front of the caregiver's body does the most work – it makes sense that there would seem to be a preference from the baby too, purely because the "passive" leg is being under-used and the active one is building up all the strength. I put passive in quotations because of course it isn't completely passive – the thigh is active but anything from the knee down may be disengaged. So, from my point of view, I believe we ourselves generally create side preferences in our babies and children. As previously mentioned, we will cover the subject in greater detail in a later chapter.

So, we know side preference is a "thing" and that it's a habit that's easy to slip into. We communicate to them non-verbally that the side we prefer to carry on is the "correct" side to be on. An awareness of this fact, making sure we do switch things up often, and active communication with the baby/child can help minimise the impact of this. It doesn't make sense to have drastically different muscle tone on one side of the body, and we'll look into this further as we move forwards in this book.

When babies do have a side preference and we're working with them to get them out of it, they won't necessarily think it's as great an idea as we do! The first sign of communication from them that they want to move to the opposite side tends to be a palpable loss of clinging strength from their thigh/s. The way this may be received by you can vary – some feel it as exactly that – a physical feeling of slight disengagement. Others experience a sudden feeling of extra weight, some an urgent need to provide more support, and many will experience some combination of these. This is a clear signal that the dominant leg (in this carry) has had enough and needs a break. Sometimes this can be rectified for a while by providing support lower down their back, meaning they don't need to cling as hard with their thighs/legs, and other times they will need a proper break, by switching sides right away.

Most of the time this first instance of communication is enough to elicit the response they need from the caregiver, but if it's not interpreted correctly (or you're mean like me and you want to see what happens if you wait a bit) the communication ramps up. My observations have shown me that there seems to be a similar process for babies and young children (verbal and non-verbal). The next step tends to be a vocalisation, accompanied by a wriggling or bouncing movement, as if to say "Hello! Are you listening to me?". This tends to be enough for most to pick up on the discomfort at some level, even if they've not made the connection that the baby/child wants to switch sides. An almost reflexive change in support, usually characterised by an extra hand or arm supporting them or a quick change from active to passive carrying tends to follow, or a deliberate switch of sides from the caregiver.

If, however, this communication is also ignored/not understood, we usually get one of 3 common escalations. These are: disgruntled disengagement from the carry (straight legs), crying (usually with wriggling or bouncing) or a physical attempt from baby/child to climb to the other side of the caregiver's body. These all of course are actions which pretty much guarantee a response from the caregiver, and the situation is promptly resolved, whether it be moving them to the other side, changing how they're supporting them, putting them down, nursing them or otherwise.

In my little experiments with Isaac, I've reached the last stage of this according to him, which is to actually physically climb across my body, and I've had reports of this from a friend too, with her then-7-month old doing the same with a little assistance. With Isaac it happened at first simply by just waiting to see what would happen, but these days (at 2.5 years old at the time of writing) it occurs very quickly as a response to my attempt to negotiate with him when he says, "I want the other side", if he really has had enough. With him though, it's harder to tell straight away if he really needs to switch or is just slipping into his preferences, which is why the climbing happens more often these days.

They also communicate by using different parts of their leg/s to push, press or squeeze. This is a bit like making us the horse to their rider. They also use their upper body and arm/s to turn, pull and push. This enables them to communicate a change in direction of the caregiver if they are trying to look at or reach something, and – as active carrying promotes interaction with the environment – it's something that tends to be picked up on fairly well. The act of physical communication with conscious response builds upon the foundations of communication already laid and can increase caregiver responsiveness.

Dyad and environment interaction

As we touched on in chapter 2.1, verbal communication factors into the carrying process more and more as language develops. It's a key part of the active carrying process. The most communicative position a baby or child can be in to foster interaction is when they are actively clinging on the hip. One of the reasons the stance of the body – partially world-facing, partially facing the carrying person – plays a big part in this. They are able to engage in social referencing as well swapping between observing the world and talking or communicating in other ways with their caregiver. They are able to see and interact with the world at the caregiver's height and are able to share this with them by being close to their face yet able to see them from an appropriate distance. They're easily able to turn into their caregiver's body to disconnect from their surroundings when they get tired or overwhelmed and have the freedom of movement, the autonomy, to choose when to clearly signal that they have had enough. By disengaging from the world, they communicate with their relaxed body that they need to switch to passive carrying. As we know, babies spend different lengths of time in different states of alertness and the ability to freely move their body helps to convey to the caregiver a change in state.

It's important to recognise that there is a difference in interaction in active carrying and passive holding/babywearing. In the babywearing industry it is frequently stated that slings and carriers are a fantastic facilitator of communication. Whilst this is true – babywearing can promote lovely communication when baby is awake and active – the way this happens is different to how it plays out in-arms. In a sling or carrier, the baby is contained. Their torso is contained, restricted up to the neck or to the armpits (which of course provides some freedom of movement for the upper torso). As well there is a containment from the top of the thigh to the knee-pit. There tends to be a lot less physical communication going on and verbal communication isn't stimulated as much from environmental factors. Babywearing and passive holding may encourage a more connective form of communication though. For example, checking in on their needs when they aren't visible (e.g. on the back), or engaging in bonding language (e.g. in a shoulder hug).

Something I love about active carrying is the way in which Isaac interacts with me if he wants to get my full attention. If he has something to tell me spontaneously (rather than me asking him questions) and I'm not looking at him already, he will either put his free hand on my outer cheek or grab both sides of my face and pull it to face him. It's wonderful to experience as even the touch of his hand to my face triggers an immediate turning of my head to look into his eyes. The use of touch to add an extra layer of communication and connection is powerful.

It's also lovely watching interactions of other dyads in active carrying. Just by observing caregivers applying active carrying principles it is easy (and fascinating!) to see the increase in communication between the pair. Whilst it's common to for the caregiver engage with others at this time, I see on a regular basis that the baby/child either attempts to divert their attention away from the third person or inserts themselves into the conversation. It also seems to be harder for the non-carrying person to resist initiating engagement with the baby due to their activity and bigger presence in-arms.

It's hard for the carrying dyad to switch off to each other when the baby is clinging. Active carrying is movement, exercise, and requires the clinger to be in a place where they have the energy to carry it out. In this place they tend to also have a need to interact with their environment and at the caregiver's side they are perfectly placed to receive emotional and practical support as they navigate this. During active carrying, babies' brains are fully engaged in learning – it's the prime time to explore environments and retain information, especially as this is most often coupled with a state of alertness where a want and need for movement is present. An active baby engages the senses of the carrying person even when they're not communicating with them. It's simply much harder to disconnect from a baby who isn't still and resting on the body.

As one practitioner of active carrying eloquently put it; when they're clinging, they are never bored! This certainly appears to be true. The entertainment is all right there where they need it, and their need for movement along with connection is also met. Babies are gaze-followers – even brand newborns[2] – and by watching us and looking to where we are looking, their awareness is raised of the environment they are in. They're born perfectly designed to follow our lead and observe and interact in their environments. Their brains are like sponges, absorbing information quickly. We see all too often that babies either tire quickly of manufactured toys or refuse them outright. They're designed to play in the way nature intended it – either on-body or exploring the world around them. When they've had enough they signal to the caregiver that they're done. This may be through things such as initiating full body contact which triggers some release of the power of their clinging, or by physically disengaging from the carry (e.g. completely straightening their legs).

Communicating needs

What implications may active carrying have on the baby's communication as they grow and develop? Obvious possibilities are an increased dialogue within the dyad (whether or not the baby can say words as we know them), a greater awareness of and understanding from the caregiver as to the baby's methods of communication, as well as more engagement with other adults and bigger people with the immediate safety of being physically attached to their caregiver. Another potential positive impact could be greater understanding of how to communicate in an effective manner. The more disconnect and/or distance from each other the greater difficulty in picking up on cues.

Babies and children tend to communicate strongly on-body when they have a need to be met. In elimination communication we feel them wriggle, squeeze their thighs together, tap the caregiver's body and so forth to communicate the need to urinate or defecate. When they're hungry their cues can be easily read, such as rooting on the caregiver's skin, reaching for the breast, sucking hands, mouthing or sucking on skin, head bobbing and moving side to side. They also communicate discomfort, and although they do this and convey many of the previously mention cues off-body too, the physical signs of discomfort can be much more easily "read" on-body. A kicking, whimpering baby off-body can be harder to read because of the way we process the information we're receiving. We're more analytical, using our higher brain more than our intuition. On-body, the brain's interpretations can be simplified, working more on an unconscious level. There is less thinking and more reacting. The feel of the body's movements (with added vocalisations at times) get across the baby's meaning in a much "louder" way. It's also harder to miss communication so close to you.

PART III

Chapter 3.1

Physiology and Anatomy of the Baby

I covered both babies' and caregivers' bodies in the first book, but I would like to revisit both in this one to explore other avenues of the physical makeup and go deeper into the science behind how babies and their caregivers are a bit like pieces of a jigsaw puzzle. They are designed to physically fit together, and this is due to things like shape, angles, and padding in certain areas.

Even when it seems that we've been made a certain way for a particular reason, there tends to be more than one reason for each design. The more I learn about how our bodies are perfectly designed for carrying and clinging, the more in awe I am, yet with every new discovery I've made about the physiological carrying process the more I *expect* to find out about our design being strongly geared towards facilitating carrying our young. I've not been disappointed, and I wonder how many more exciting discoveries there are ahead of us!

To go beyond the reflexes and voluntary actions to understand the carrying process, we must explore many parts of a baby's physiology and anatomy. Some of this will be left until chapter 4.3, as the senses involved in carrying require separate focus.

Central nervous system (CNS) and neurodevelopment

We touched on the CNS in chapter 1.3, and now we're going to further explore its function in carrying. There are many ways in which carrying supports neurodevelopment, some of which, again, we'll explore further on in the book. For now, let's look at the connection between physical movement and neural development. Basic movement patterns are essential for normal neurodevelopment[1] and missing stages of development can interfere with the body's normal functioning. Behaviours are learned by repetition – good or bad – and whether or not a baby/child learns clinging behaviours plays a part in the default pattern established through countless movements over periods of time. These movement patterns go on to define a child's posture amongst other things. Seeing the effects of behaviours is very difficult at a young age due to the time it takes to reach a point where negative effects are visible from the outside of the body.[2]

Neurodevelopmental movements in infancy are the spontaneous movements babies make, often linked to the primitive reflexes. Unrestricted movement is essential for the normal development of the nervous system, brain and body. Primitive reflexes which haven't integrated can cause learning delays[3] and other issues for children and adults.[4] These retained reflexes, once identified, are able to be integrated at any age in life with specific exercises[5] but it's in a person's best interest to not have the issue in the first place. This is why it's important to give babies and children the opportunities to interact regularly with the world around them and to use "baby containers" with care.

Sensory and motor neurones are fired off by repetitive movement, and *clinging is movement*. Babies are preadapted to cling to their caregiver/s and various primitive reflexes are triggered almost constantly or intermittently in carrying. These bring into question whether or not we may find that missing part or all of the on-body clinging developmental process affects neurodevelopment in some way. On the flip side, completing this process may have many perceived "benefits", which are simply basic developmental achievements. It's so important for babies and children to be given the opportunity to engage in movements which are essential for a healthy nervous system. Without sufficient access to normal movements they're more likely to need to take on adaptive patterns to compensate.[6]

How the CNS works and matures is partly to do with the spinal cord sending sensory and motor messages – which are received via nerves – to the brain and the rest of the body. The brain then sends motor commands through the spinal cord and movement is generated. The spinal cord is responsible for both reflexive and voluntary movement. The brain stem connects the spinal cord to the brain and controls reflexive and involuntary actions. The cerebellum assists the coordination of voluntary movements of the muscle, maintains muscle tone and receives and integrates the sensory messages sent through the nerve pathways. The cerebrum perceives, processes and integrates all sensory and motor information. It controls things like maintaining upright posture and contains centres which create new patterns of movement. The frontal lobe contains the primary motor cortex which controls voluntary movement.

As we know, babies and children go through significant development of the brain during infancy and childhood. One of the initial parts of development outside the womb is focused on body control. In carrying, the CNS processes sensory stimulation such as the level of support the caregiver is providing or changes in balance and fires back with an appropriate response. The infant makes reflexive responses at first, which gradually develop into voluntary reactions (with some reflexive actions still in place throughout life) in response to the varying input.

Cardiovascular and respiratory systems

An interesting study focused on the calming response of carrying.[7] The biggest effect they found was the impact of the act of standing up from seated (going from "holding" to "carrying"). The heartrate of the babies was affected most in those who were crying during the holding phase, but a calming response was still observed in the ones who weren't crying. They also found that babies' general movements also stilled during this time. A response wasn't observed in infants who were sleeping during both phases.

This is very interesting beyond the obvious conclusions that standing up and moving calms a baby better than sitting down and holding them does. Being a helpful participant in the carrying process requires attention and controlled limbs – the calming effect likely promotes focus. Another benefit is that it would conserve energy. Does this mean that – as well as infants having a built-in need and expectation to be carried – they have an adaptation which can make carrying easier for the dyad?

Clinging is, of course, a form of exercise. Babies are designed to carry out movement tasks which don't put undue stress on the cardiovascular and respiratory systems. The nature of the movement of carrying can be compared to a relatively fit and healthy person going for a stroll. Breathing and heartrate stay within a normal range and conversation can easily be held. Active carrying is not stressful to the baby and is a gentle endurance task.

Head size

A newborn baby's head makes up roughly ¼ of its total body proportion, whereas an adult's accounts for approximately $1/7^{th}$. This makes for difficulty in holding the head up, especially with underdeveloped neck muscles, and doesn't seem to make much sense as it's a hindrance to carrying, however, it is required because of the size of the advanced human brain. Could this trade-off have any hidden benefits? It does seem to tie in well with the other stunted areas of development at birth and restricts them in an appropriate way as they acclimatise to the world around them. As head control is gained their sight is improving too. By the time they engage in active clinging their proportions have changed somewhat and control of the head, neck and park of the trunk has been achieved. Another way in which this may aid humans is the way it inconveniences the caregiver. Although this seems like a negative thing, we know that it's traditional to rest for the first 4-6 weeks postpartum. Having a very active baby is not conducive to this. Even if the caregiver needs to move quickly it's not a huge obstacle as the baby is small enough to easily be supported and secured without needing to stabilise its own head. We will look at another benefit of having a seemingly large and heavy head in the following chapter.

Upper body

In the early weeks, as we have seen, active carrying has a primary focus on upper body development due to the sequential nature of spinal development. It's all about developing the cervical curve of the spine, building up strength and gaining control of the upper extremities and torso. The way in which this development plays out seems to be beautifully synchronised. As upper-body control is practiced and reached, lower body clinging reflexes are still in full swing. The focus is only ever on one area of development, rather than the whole. The lower extremities utilise their reflexive properties until the upper extremities and torso reach a point of appropriate development and good control.

It's interesting when we observe the cultural trend towards passive carrying. The upper body of the child tends to receive less negative impact than the lower body. Passive carrying of older children ranges from semi-active to passive. Semi-active carrying usually involves full support of the lower extremities and freedom of the upper body. This means the upper half of the baby's body gets to work more, developing strength in carrying in a closer-to-normal way. Even when the lower half of the body is being supported we tend to see an active engagement from the baby/child to the caregiver's arm or shoulder in passive carrying. They make contact in some way which provides an anchoring point, stabilising them whilst they use their core to a certain extent to balance and counter movement. This is great news, considering the trend towards passive carrying. The whole body need not be influenced so negatively.

The young baby's arm is of a sufficient length to interact well with the caregiver's upper arm by the time they're developmentally ready to transition to the hip. The way the arm bends at the elbow means that they're able to create a stable support. Hands are also able to grasp onto clothing and as development continues there will be variation between the way they use it. Sometimes grasping will be used to aid clinging and at other times an open hand resting on the arm – or none at all – will suffice.

Spine

So, we touched briefly on some of how spinal development occurs when we looked at the tonic labyrinthine reflex, but now let's look at some other aspects. Newborn babies look as though they sit with their spine in a "C" shape on-body, and this description is useful as a visual aid in understanding positioning. However, their spines are much straighter than they appear to be from the outside[8] and this is where we need to be mindful of our language if we're wanting to understand how carrying works at a deeper level. This straighter spine in the thoracic and lumbar regions seems ideally suited to carrying, as the caregiver's forearm rests gently against their upper, mid or lower back when supporting clinging. Although the arm doesn't behave in a way in which compromises the spine (e.g. pulling baby to their body), the ways in which the baby moves and uses it for support are more suited to a straighter spine. Even when the lumbar curve is in place it doesn't impact on the shape of the spine in clinging as the spine is neutral in aligned sitting. Even so, clinging at this point should be advanced enough that support of the lumbar spine is rarely needed.

It's worth considering the initial shape of the spine versus that of the crawling and walking child. Before a baby is able to do these things, their lumbar spine is relatively flat. This shape is conducive to active carrying support on the hip. A convex spine would be detrimental as it places the carrying-person's arm at an angle. As they begin to develop the lumbar curve the support ventures upwards in dyads practicing active carrying principles. This means that the newly strengthened spine can take on its role of actively enabling strength and support of the lower body. The caregiver's arm moves up to provide a bracing point for the already-strengthened upper body, enabling the lower body to take on a more active role. This change in roles seems to be directly linked to the developmental progress of the lower body. The lower body is given even more freedom to work, move and develop once independent sitting is achieved. This freedom is instrumental for reaching the developmental milestones of cruising, standing and walking.

Protection of the spine from the movements of the caregiver is also needed, and the way the spine is shaped in a younger baby, and older ones when in a seated position, allows the forces exerted in the movement of carrying to be distributed effectively. This is also why we see a reflexive change in spine and hip position in young babies through the TLR – babies this young cannot voluntarily protect their spine so the reflex does it for them. There's a lot to be learned from how the spine straightens in relation to the hips. In active carrying the it's straighter and the leg position is either chair-like or close to it. In passive carrying the spine is relaxed and the legs come up higher in hip carrying. This trend continues through the carrying developmental process, and we see that as older babies develop and get heavier they tend to cling at a lesser angle degree than younger ones.

The development of spinal curves is required to enable babies to progress from needing to be carried everywhere to being able to explore the world independently. Active carrying supports spinal development by allowing the practice of movements needed to complete each curve. As we looked at before, the initial active shoulder hug allows for the practice of head and neck control, which becomes a permanent feature once the cervical spinal curve is fully developed. The lumbar curve which develops through supporting the upper body whilst prone, crawling, creeping and walking is also aided on-body by the previously mentioned reduction in squat angle and subsequent straightening of the spine.

There is (rightly) a big focus on respecting spinal development in the world of babywearing. However, in some parts of the community there can be a lack of understanding as to how this development occurs and ways in which it is impacted. Spines are moveable, even in the newborn, and the appearance of the cervical and lumbar curves does not happen overnight, as we learned from the information about how the reflexes affect development. Whilst it must be stressed that *forcing* it into extension is definitely to be avoided, there are many ways in which the newborn and young baby's spine is effectively straightened out each and every day. Every time they are placed on their back, it is flattened to some extent - especially when they're sleeping, and muscle tone is suppressed. That accounts for over half of their positioning every single day! Each time they're carried in-arms the movements made (especially of the head) create flexor and extensor responses. With every episode of "tummy time" their spine is moved into a certain degree of extension. A baby who is able to support their upper body with their arms will be creating an even greater lumbar curve, yet this happens long before this curve is properly developed. The point is, infant spines are meant to move and to exercise in a manner in which helps prepare them for future stages of development. It's important that this occurs in a developmentally appropriate way, of course.

When supported appropriately on-body, babies are able to achieve positions they would otherwise be unable to off-body. For example, even a newborn baby may be supported in a suitable manner facing outwards in-arms, without compromising the spine. This is due to the way the caregiver's arms work and are able to instinctively adapt in ways devices cannot. Similarly, in a correctly supportive baby carrier the older baby is able to be held in a way which keeps the spine in a position appropriate to the level of development of the pre-independent sitter who has good upper-torso control. The issue is how the baby is *supported* rather than a position being inherently bad. For example, the upright carrying position used to be perceived as harmful to babies' spines in Germany at one time, but it was shown that this is not the case.[9] This is an example of why it can be so beneficial to look at the positions we naturally use in-arms as well as having scientific research to back it up.

The opportunities for spinal development presented to the baby during active carrying are significant. Young babies are designed to spend much of their waking hours on-body, and it could be argued that they're also designed to sleep there too.[10] Therefore, it wouldn't be useful for their development if they were created to be mainly carried passively or if in-arms carrying was inherently harmful. The initial regular movement of the head and neck in the shoulder hug and chest-to-chest positions of course promote cervical spine development and once good control is achieved the thoracic spine receives the main focus. Infants receive a different sort of spinal support in this type of carrying. Much of the time it's simply touch, which is moveable, and it serves to support the head, neck and/or back.

When upper body control is achieved, and the baby moves to the hip, support moves down the spine and the nature of support changes. This is due to clinging – with no support of the legs the primary support comes from the forearm across the spine rather than a hand. The torso is doing less work than the legs in active hip carrying so requires greater support. However, the work of the legs affects the spine and produces movement and aids positioning. Also, support in active carrying is not full support. As we've seen with the other stages, the type of support in active carrying is merely a stopping point. It allows for movement and practice of postural control. The same is so for the lower spine. A buffer is provided so that the baby has a clear boundary in its clinging environment and is just that – a boundary. As we've covered previously, active carrying in no way relies on the caregiver pulling the baby/child to their body. They simply define the outer limits of the boundaries of carrying. This means spinal movement and development is fostered in active carrying, even though outside visual interpretation may suggest otherwise.

We know that freedom of movement is a key factor in reaching physical developmental goals. Therefore, we enable movement of the area in which the main focus of development is on, coupled with support nearby, which can be moved upwards or downwards as needed. Thankfully it is fairly simple to feel and observe what the baby needs. There's no pushing them to work beyond their capabilities. We also know that a combination of experiences in the world around them helps babies develop in specific ways. For example, spending time on their front at an age where they're able to support their upper torso elicits extensor postural reflexes which help prepare their body for crawling. This suggests that the influence of other developmental processes may influence carrying behaviours too. Simplistically, it means that spinal development is happening in many different ways and is likely facilitated on-body too.

Every single stage of physical development relies on movement, loadbearing and interaction with the environment around them. This is facilitated on-body by allowing them the opportunities to develop their clinging strength, which develops leg muscles, which aids in the transition from crawling to walking. There's also a transition of forces from sitting to crawling to walking. What do you think provides the bridge between these phases? Carrying, of course. It would be interesting to find out how much smoother the learning of each of these developmental processes is when active carrying is regularly involved in the infant's day-to-day environment.

Hips

A newborn baby's pelvis is made of mainly cartilage and the process of ossification begins soon after a term baby is born. The period of rapid development ends by age 4.[11] It's important for the natural range of motion to be respected so that good hip health is continued. This range is specific to the infant as there are differences between individuals, varying in tightness of the ligaments and individual ranges of motion. There have been various studies conducted concerning optimal hip angles in babies, so let's look at some of these and see how they may apply to in-arms carrying.

Büschelberger researched hip angles of the newborn baby. He found that a 40° angle of abduction (spread) and a adduction (flexion/squat) angle of greater than 100° was optimal for hip health.[12] Fettweis says that adduction of the hip should never be less than 90° as it will cause the femoral head to press against the roof of the hip socket and may cause damage to it. He acknowledges that this is due to the way weight is borne into the joints in babywearing.[13] Kirkilionis studied the natural positions babies sit on the caregiver's hip when the baby is supported in a way that allows their legs to cling freely. She found an average of 45° abduction and 90-120° flexion occurred. She also noted that the angle of abduction increased the further towards the front of the caregiver's body the baby sat, with 56° being the highest angle recorded.[14] Interestingly, Fettweis' thoughts on maximum abduction is that it should never exceed 55°, which is just 1 degree less that the highest angle Kirkilionis saw in active clinging.

Average angles have been recorded based on different studies of the range of motion (ROM) of the hips in infants, such as the research referenced above, yet the information which piques my interest most from these and other studies is the fact that the normal ROM vary greatly from infant to infant.[15] This, along with the fact that hip joints are required to move and bear weight in many different ways, suggests that definitive "best angles" may be questionable. For example, even though babies have a squat reflex it isn't activated all of the time. Babies lay with legs down too; the longest periods of time being when they are asleep. The big focus on hip health and optimal angles comes from advice to do with the much-researched condition of developmental dysplasia of the hip (DDH). With this in mind, we're going to first explore this before coming back to the topic of in-arms carrying.

DDH is a condition which develops around the time of birth, after birth or during childhood of which the exact cause is unknown.[16] Screening is completed within the first 72hrs post-birth in England, and a second check is offered at 6-8 weeks old. The International Hip Dysplasia Institute (IHDI) and other sources such as the NHS state that DDH cannot be prevented.[17] Therefore, the real caution should be focused on positioning where DDH is present rather than applying medical advice to babies with healthy hips. In England there are an estimated 1 or 2 cases in every 1000 live births that need treating.[18] That's 0.1-0.2% of babies. It's also estimated that 8/10 of DDH cases are found in female babies.[19] That means 20% of this 0.1-0.2% are male. 6/10 cases are found in firstborn children and is 12 times more likely in babies where there is a family history.[20] The hips have developed sufficiently by the age of 6 months to be at lower risk of DDH.[21]

"The most unhealthy position for the hips during infancy is when the legs are held in extension with the hips and knees straight and the legs brought together, which is the opposite of the f[o]etal position. The risk to the hips is greater when this unhealthy position is maintained for a long time. Healthy hip positioning avoids positions that may cause or contribute to development of hip dysplasia or dislocation. The healthiest position for the hips is for the hips to fall or spread (naturally) apart to the side, with the thighs supported and the hips and knees bent. This position has been called the jockey position, straddle position, frog position, spread-squat position or human position. Free movement of the hips without forcing

them together promotes natural hip development." – **IHDI website statement**

So, the cause of DDH is unknown but it is believed that certain positions could contribute to it. The IHDI states that legs held in extension and brought together is the unhealthiest and riskiest position, especially when it's adopted for long periods of time, and that the spread-squat is the best. It would be natural to assume that the squat position should be aimed for as much as possible. However, they also state that free movement of the hips promotes natural development. This suggests that static positioning may not be desirable. Indeed, if we look at the ways in which young babies move and interact with their environment, we see them cycle through their normal range of motion regularly. For example, they kick their legs into extension, are placed on their front to play (hip extension) and lay flat or hold their legs in a squat position. The IHDI does not advise against "tummy time" which appears to be closest to the most undesirable position of hip extension and is used for increasing periods of time in the first 6 months of life. No, it advises against forcing/holding unnatural positions. This tends to be more of an issue when using devices such as baby carriers as the infant is held in a static position and for longer periods of time than if in-arms.

Something about in-arms carrying which may be considered controversial is the fact that it doesn't always happen in the spread-squat position which is recommended by the IHDI and favoured in the world of "ergonomic" babywearing. The legs can be at sub-90° in the shoulder hug, for example. We must remember to separate babywearing and in-arms carrying as different practices and recognise that we cannot simply transfer all in-arms "rules" to babywearing and vice-versa. Similarly, in the most common form of active carrying - hip clinging - the legs and hips don't tend to be symmetrical. Positioning also varies depending on how far to the front of the body they are and the dominant vs. non-dominant legs' engagement with the caregiver's body.

It must be highlighted that data used to support the notion of babies needing to be supported knee-to-knee in a knees-higher-than-bum position is not based on sling and carrier use. The only on-body positioning research the aforementioned study from Dr. Kirkilionis, focused on the in-arms hip position. This was looking at the position they adopt rather than how it impacts on the hip joints. To date, no research has been conducted on the use of "ergonomic" carriers, narrow-based European carriers or any specific positioning in these devices or other slings. Nothing has been studied in relation to how different angles impact the hip during in-arms carrying either. It is imperative that research is conducted specific to each of these situations and positions before we state with authority what is good or bad for developing hips. It is also of the utmost importance that we remember that loadbearing is different in various positions and devices, and the difference are even greater when comparing active clinging and sitting in a baby carrier. It is certainly helpful to use studies focused on certain areas pertaining to the subject we wish to back up with scientific research in terms on erring on the side of caution if a caregiver chooses to do so, but awareness is needed.

If certain positions were inherently harmful for the developing hip we would expect to see a much higher incidence of DDH. Babies have been held in narrow-based carriers for decades, and in shoulder hugs for indeterminably longer. Passive positioning is highly questionable in terms of what sort of pressure is being placed on the hip joints. Some babies and children do not adopt any sort of seated position when picked up and are pulled into the caregiver's body by a forearm on the thighs of straight legs. Other babies are held in a high, off-centre hold which presses a straighter leg against the body and the outer leg presses against the inner. The reality is that 99.8-99.9% of babies do not develop DDH, and the hip is one of the most stable of the synovial joints. Another thing to remember is that DDH is a medical condition. The normal hip joint has properly functioning ligaments, as opposed to hips with sublaxation or dislocation. For babies with this condition, they are more susceptible to the effects of positioning. The advice adhered to is based on using the guidance for medical management of this condition, transferring this to healthy babies. This isn't to say that this should be ignored (especially in the case of passive carrying and babywearing, where the load is static), just that it cannot be applied to all positions for the hip-healthy child.

Again, it's incredibly important to acknowledge the different ways in which weight bears on the hip joint in different positions. It's also pertinent to be educated in the ways in which the hip develops to sustain various load-bearing activities at different stages of development. It is not as simple as saying a narrow-based carrier or an in-arms position where the legs are below 90° is bad for the baby. From the prescriptive advice regarding optimum hip position, it could be easy to conclude that a baby should stay in a static position for as much time as possible, so as to develop the healthiest of hips. Of course, this is not being advised by any organisation, medical or otherwise. Until we have concrete evidence of certain positions being inherently harmful, looking to the normal development of a healthy baby can provides clues as to what may be natural in terms of positioning and load-bearing.

So, back to the shoulder hug position. This is used frequently in the early weeks and months, and babies don't tend to sit in a squat position past the first month or so, especially when active in the carry. In fact, in both the active and passive shoulder hug variations, their knees can be lower than the hips and, as we know, research into hip dysplasia has suggested that this is less than optimal for babies. It must be noted here that the true angle of the hips in a shoulder hug can be hard to judge with the naked eye if there is a degree of forward-tilt to their torso – the hip angle does tend to be higher than it may seem in this position.

Whilst some in-arms positions may be sub-90°, naturally supported positioning doesn't pull legs straight down or together. Loadbearing comes from sitting on the forearm. As we know, babies have to weight-bear in different ways as they learn to travel independently, and so their joints gradually do so at different angles all the way to the hips being in regular full extension during standing and walking. Before they reach the stage of crawling, their joints bear weight in other ways such as supported standing happening for short periods of time from when a young baby indicates they want to "stand" or bounce on a caregiver's lap. Being carried is an example of longer periods of loadbearing. It seems unlikely that natural in-arms positioning such as the shoulder hug would be inherently harmful.

If just the bum is being supported, the loadbearing happens in a different manner to, say, sitting on the forearm. Either the legs rest passively, dangling down, or they cling in a 90° angle or higher. There is also a difference between dangling legs in baby carriers and in-arms. Carriers with a narrow base are supporting the baby in a different manner to arms. Pressure is created in the hip joints from the weight of the baby going into the seat of the carrier and the edges of the material being either side of the crotch. In-arms, there is no additional pressure on the joints when supporting just the bottom. When no support is provided for the legs or bottom the load-bearing happens in yet another way. When we're standing still the load is from the actual weight of the baby's femur angled into the acetabulum. When we walk there is the addition of movement creating variable pressure. Keeping with the idea that carrying is movement, clinging helps provide good blood supply to the hip joint. The head of the femur in the hip socket moves every time the caregiver and child do, through walking and positional adjustments. The body expects and needs movement. Although there is some movement in passive holds, we see much more in active carrying.

Babies also behave differently in-arms to each other. Some are very straight, with the squat position being near impossible to evoke on-body when awake and active, leaving them held at less than 90 degrees in passive holds and hitting 90 degrees or less in clinging positions. Does this mean they're harming their joints with less variation of angles going on? Some babies just naturally have higher tone but for others this may be caused by other conditions, including colic or reflux. It would be interesting to find out what impact this may have on the joints, if any.

We saw in chapter 2.2 that babies stretch out and lose their states of reflexive flexion as they develop. As with spinal development, this happens with regular movements rather than suddenly appearing. Babies are regularly stretching and kicking when active and alert off-body and adopt a constant straighter position when at rest or sleeping, when the squat reflex isn't invoked. Whilst hip flexion contracture may still be seen in hip-healthy babies even at 6 months this is of an average angle of just 19° at 6 weeks and 7° at 3 months.[22] We know that the hip joint is designed to weight-bear in different ways at different stages of physical development. By referring to other physical developmental processes it seems reasonable to speculate that the natural positions babies adopt in-arms when we support them appropriately would be good for their bodies. The human body is designed incredibly well. If we're meant to do something, then our bodies are given the abilities to do so without causing harm. The more difficult thing to work out is what is the most appropriate way to support them, as we are a catalyst in this carrying environment. There most certainly are ways in which we carry children that seem to be questionable from a hip-health point of view. For example, the compression of the legs and hips in a high, off-centre hold where the child's legs are pressed to the caregiver's body, one leg higher than the other, or holding babies facing outwards by using a hand to bear baby's weight at the crotch. With all this in mind I believe it is of great importance for the impact of supporting normal carrying angles to be studied, especially the support of babies in sub-90° angles in the shoulder hug.

The last thing to remember about hips is how the function of the leg and hip evolves as the baby develops. If we go back to the information related to crawling onwards in "Developmental Process" we saw that the hip joints are further forwards[23] meaning the range of motion of their legs favours the front of their body. Even when the hips move in alignment with the spine backwards motion is limited, and this is particularly noticeable in the gait of new-walkers and young children. The transition to a full range of motion of the legs as an older child is due to the increasing lordotic areas of the spine as development continues. Babies' primary method of movement (from one place to another) is by being carried in the early weeks and months and as they build up muscular strength they begin to move in other ways which encourage the hips to be aligned in a different way as the spinal curves develop. This is yet another way in which babies' bodies are designed to evolve the nature of their clinging as they grow.

Body fat

There is a big question mark surrounding the subject of infant body fat. Along with the significantly greater weight of the human infant when compared to non-human primates, there's the matter of a marked increase in body fat percentage. One study found an average of 13.3% at birth, 24.5% at 2 months of age and 31.2% at 4 months.[24] Why are babies so well-insulated when the makeup of their milk is low-fat, and they're designed to stay attached to their caregiver and feed frequently? There are several theories about this, such as the large size of the infants brain accounting for over 50% of their daily basal metabolic rate,[25] a protective factor in case the caregiver has to be away for longer than usual or dies, or additional insulation, for example.

The brain's energetic costs don't seem to support this theory as the fat stores increase as the brain grows, rather than being maintained or depleted; they're not tapped into to sustain the organ. The protective factor seems to be a plausible explanation until we realise that all other non-human primates haven't got this special feature. Why would just we? Insulation seems plausible, get babies are designed to be in close contact with their caregivers – being clinging young – and the caregiver's body generates heat which can be shared. Also, why is the fat not gained *after* birth, seeing as they must already make their way through a small pelvic opening? Surely it would make sense for them to be as skinny as possible to make that journey easier?

However, if we look at fat through the lens of in-arms carrying, it appears to play a very important role. Fat is cushioning, and human clinging creates more pressure due to the specific nature of it. We don't use our hands and feet as the primary attachment mechanisms. Human babies' hands are poorly adapted to independently cling for long periods, even if their caregiver's body had the long hair needed for such a thing. Fat also works with the pliable nature of skin to create a mouldable, slightly moveable surface to cling to. Even if we aren't doing much carrying right away, we're doing a lot of holding and in the coming weeks active clinging is not yet happening. The pressure of their weight on-body is present and the mouldability of the surface is still preferable. As the baby grows it is useful to have additional fat to cushion the increasing weight whilst clinging abilities go through the main part of their development. As active carrying first requires the clinging reflexes it is less developed. The squatting angle is greater as the baby is more "slotted on" the body. As voluntary clinging emerges and develops, the reliance on fat for cushioning decreases as the drag is less. As activity levels increase the body fat decreases[26] and this is correlating with the strengthening of leg muscles.

How babies' muscles develop

In a culture where baby "containers" are frequently marketed at caregivers, coupled with the ever-persistent parenting advice to not hold babies for too long to encourage "independence", babies tend to receive much less freedom of movement and time in-arms than is biologically expected by their bodies. Babies laying flat or reclined are in a position of rest – they're sedentary. Their core is disengaged, and their weight is on the back of their pelvis, head, shoulders etc. Babies and children are designed to *move*, as are we. They don't tend to stay still for very long and this is completely normal. Yet societal expectations encourage stillness and sedentary behaviour, so discouragement of movement is drilled into us from infancy.

We know that using our muscles is what enables development and strength. If we don't use them they atrophy and the more they are used, the more they develop. It's important for me to reiterate here that everything I'm speaking of is to do with *normal* development. This is not encouragement to push a baby into achieving some sort of athletic feat, yet the capabilities of a normally developing body can seem astounding in societies where we've moved so far backwards in our quest for convenience.

Specific muscle fibres begin to appear around 30 weeks gestation. Type I fibres will account for approximately 40% of muscle at birth, type II a and b for 45% and type II c (undifferentiated) for 15%. Type I is very resistant to fatigue, capable of producing repeated low-level contractions by producing large amounts of adenosine triphosphate (ATP) which generates energy for muscle contraction. Type I muscle fibres are capable of producing lower amounts of ATP over longer periods of time and rely on oxygen to do so. Types of muscles containing mostly Type I fibres are postural muscles such as the neck and spine. Type II a produce fast, strong muscle contractions and fatigue faster than Type I. If a greater clinging capacity is required (one which goes beyond the baby's current levels of endurance) the other muscle fibres are recruited to help out. This results in faster rates of fatigue, as per the example in Chapter 1.2. Clingers require the development of slow-twitch (Type I) muscle fibres to make long periods of active carrying possible. These fibres are responsible for maintaining endurance, and clinging is obviously an endurance activity. It would be a hinderance if clinging made their muscles ache, which tends to occur in activities which rely on Type II fibres.

As with other areas of human design we have a ratio of muscle fibres specific to us. Some people have more slow-twitch and others more fast-twitch (Types II a and b), and we have a range in which we can achieve peak individual performance. We would expect the endurance needed for clinging to fall within a "normal" percentage of slow-twitch muscle makeup, but this is something which needs looking into further. It is clear, however, that genetic predisposition may impact on clinging abilities. When carrying or observing others carrying babies and children with weaker muscle tone, for example, it's clear the challenges presented to the dyad. Whether this is something that can be improved upon will depend on the degree of hypotonia and if there are other medical issues present. On the other hand, we also see some cases of children who weren't given regular opportunities to cling to the caregiver, yet when offered the chance they either relearn exceptionally quickly or can pull off clinging on-body without instruction. It appears some have a lesser or greater advantage than the average baby.

So, to encourage the normal development of the lower body caregivers will ideally support clinging on-body. We're talking about clinging with the thighs, plus the calves, ankles and feet. Of course, full-leg clinging is not always seen, usually due to the clothes being worn. Skin-to-skin clinging is where we usually see the most amount of leg and foot contact and activation. Muscle development is another reason for switching up how caregivers carry, to try to avoid a strong side-preference developing. We know that in hip carrying there is a dominant leg, which does much more of the work. The non-dominant leg (to the back of the caregiver's body) will develop at a different rate if a side-preference is indulged, and the difference in leg strength and endurance will be easily noticeable in carrying.

As we know, in reflexive carrying the baby is able to hold the spread-squat position for incredibly long periods of time. This muscle activation is obviously very useful for an infant whose legs will not have primary developmental focus for many months. On-body the use of this position is combined with the clinging reflex which sees them activate their muscles further. As babies start to lose their reflexive squatting and clinging behaviours they must start voluntarily reproducing these actions and build up the strength and endurance to carry them out for longer periods of time at will. We know how much harder it is to adopt this position without reflex as we can test this ourselves by laying and adopting the position and trying to hold it. That's just static without factoring in the clinging required to hold it on-body. If active carrying principles are not applied there's little to encourage them to be active participants, endurance in clinging on-body is not developed, and so begins another stage of unconsciously encouraging sedentary behaviour. It would be helpful to find out how the muscles impacted when used for clinging as well as in other developmental processes. Does clinging on-body develop the muscles in certain ways which aid climbing and clinging off-body?

It may not seem that obvious how much work their arm does when a baby/child is actively clinging in a hip carry, especially as it tends to look as though they're just resting it on the caregiver's arm. In fact, it's doing a great deal of work and this is very much noticeable to the person carrying if the child takes it away for whatever reason (e.g. needing two hands to hold something). The active arm in carrying doesn't just act as a stabiliser, it provides support for the upper body. The arm is receiving movement, it's actively engaged (not as much as the legs) and is very much part of the carrying process. Again, this is something to think about in terms of muscular development and varying your carry to promote development of both sides.

Active carrying is also incredibly beneficial for developing a strong core, as it's needed to make clinging possible and to keep the upper body stable when movement occurs. A strong core is important for many things – alignment, stabilisation, movement, as well as achieving developmental milestones. Active carrying is one of the earliest forms of movement outside of the womb. Developing this into clinging and continuing this form of carrying is simply the biological norm. Babies' bodies expect it. They're preadapted to participate in carrying and this helps with normal muscular development.

Legs

Something of a bemusing nature is the bow-leggedness of babies. They need to be able to fit in the womb, and the position they end up in is said to be the cause of the slight genu varum. Mild compression of the medial growth plates of the knees during load bearing (standing and walking) results in faster growth of the bone on this side, which effectively corrects the condition through the first years of walking.[27] During this time of bandy-leggedness, babies have a space between their knees and the space between the thighs is greater than when their legs have straightened out. This shape helps them to fit well to the caregiver's body. The inward curve of the tibia allows for easier clinging of the lower leg in hip carrying before they're sufficiently developed to hold on tight enough with a straighter leg. Their body fat must also be remembered. Fat is soft and pliable. Compression of this also frees up more space between the legs and helps mould them to the caregiver's body.

The height of the baby in a hip carry puts the lower leg at a position to connect with the caregiver's lower abdomen. If this is the postnatal parent, then it's likely to be somewhat rounded from pregnancy for many months yet. This accommodating space along with the flexibility of the ankle joint to orientate the foot towards the midline promotes contact with the caregiver's body. In a chest-to-chest hold the legs again conveniently rest around the abdomen, with the curved shape again fitting nicely to the body. With a toddler, the residual curve in the lower legs provides the shape suited to cling to the curves of the hips in a front or back carry. As they straighten out over the second year of life, their legs usually reach a length where the curved shape is much less needed. The thighbone length places the lower leg on the outer side of the caregiver's body and the way they cling changes.

Bone adapts to the mechanical demands placed on it and is obviously more susceptible to change when babies are younger. They're born with the weaker, more flexible woven bone which is then actively replaced by the more mature lamellar bone from around one month old.[28] Bone is designed to be loaded in multiple ways, and the activities the baby partakes in affects how it matures. The leg bones increase their loadbearing as development advances, only taking on the full load of the body once assisted standing commences. By being clinging young their legs are protected from other forms of loadbearing they're not yet suited to. As they are carried less and can transport themselves independently more their legs adapt to be better suited for bipedalism, and so the clinging adaptations of the leg disappear.

Feet

Babies' feet work in a unique way in carrying. When they're being used to cling to the caregiver in active hip carrying they curve inward and the toes also participate. This is mainly seen in skin-to-skin carrying and is seen more in skin-to-clothing (babies' bare skin) than in clothing-to-clothing. Feet have a much better range of motion for internal rotation when compared to external rotation, just like the hands. The fact that internal rotation helps improve clinging in carrying yet is still retained (albeit at a lesser extent) once the carrying years are over suggests that we are still designed to do *some* clinging past this time. The legs grow out of their adaptation for clinging as it would hinder walking. Notice that as the tibia straightens the knees also lose their slight flexion in walking and the child's gait develops further. Internal rotation isn't activated during walking therefore can be retained.

Babies have fat pads in their arches, which at first thought seems counter-productive to carrying as a shaped instep and prominent heel and sole would surely better work with the caregiver's body, right? Leg and foot muscular development of babies and children is not advanced enough to support feet with fully developed arches as babies. So, if this is the case, how are their feet designed to support clinging beyond the reflexes of the foot and ankle? It appears that they're adapted to work well in carrying via the internal rotation and reflexes of the foot rather than their shape. Instead, it may be that these things could aid foot development via the process of clinging.

The feet contain no bones at birth; just cartilage. It takes around 18 years for the foot to develop into its adult structure so how they're treated during the childhood years can impact greatly on foot health. What we put on growing feet and what activities babies and children do with bare feet affect their structural makeup. Shoes impact on foot mobility in general[29] so it's understandable that the activity of feet in carrying changes when babies and children are shod. Whilst this may come as a surprise, even socks can affect carrying. Every layer is a barrier to sensory input, as we will explore later on. Socks and babygros can impact on the activity of the foot but soft leather and more structured shoes do so more, with structured shoes having the most effect. When foot coverings are needed in carrying (to protect from the elements, for example) soft, flexible ones are ideal, or if this can't be avoided the caregiver can look to the clothing materials being used to ensure clinging isn't being further impeded.

I would be particularly interested in seeing how foot development may vary, and if there is any difference between a child's feet where the caregiver only actively carried them on one side of the body. Two questions I have are: Does clinging with the foot impact on arch development in any way?

Does it affect range of motion of the ankle joint? I would be very interested to find out the answers.

Chapter 3.2

Differences Between Caregivers' Bodies

Carrying absolutely differs between men and women, and biologically male and female bodies. Within each one there are again other variations and spectrums, so we will just take a look at your average biologically female and male bodies rather than going deeper into the nuances of inbuilt and societal carrying patterns. Here we will explore how different body structures and shapes affect different carrying positions and create a better picture of how this may affect the differences between how different people carry. Starting at a biological baseline opens up lots of opportunities to work out how to adapt to different parenting setups and so forth.

Upper body

In a passive shoulder hug the baby is put high on the caregiver's body. Their head is supported in the beautifully designed space between the neck and the point just before the acromioclavicular joint, which is padded nicely with the trapezius muscle, providing a sort of "pillow" as well as a slight hollow for the newborn's head to fit into. This also gives the caregiver help with a heavier perceived load as the infant's heavy head - a whole quarter of their body size at birth – is borne by a large muscle (plus surrounding muscle groups) and the forces of that weight are going down through the caregiver's body rather than the arm being required to support all the passive load. This is a fantastic design to aid carrying before the baby is developmentally able to participate in active clinging.

The clavicle is a bone which varies greatly in size and shape between individuals. It's an "S" shaped bone which seems well-shaped for passive baby carrying. A study showed left clavicle bones to be longer and more angled than right ones[1] which is interesting if you think about studies showing a majority left-side preference in carrying for women and right head turning preference of babies. These are discussed and referenced in the next chapter. It also showed that left clavicles and those of the female body were straighter than male and right clavicles. Clavicle angles, though, showed that for left clavicles there were less biological differences than right ones. Right lateral and total angles were significantly greater in females. The way in which the clavicle is shaped and angled affects the "nestling" space by the neck. One reason for asymmetry could be to do with the differences in loads placed on each side.[2] Interestingly, females clavicles grow at a much faster rate than males.[3] This could present another biological link for females being expected to be primary caregivers.

On a female body, the angle of the baby's body is slightly diagonal but curved to the upper body, around the breast, with knees nestled around or underneath. This supports the spine at rest well and suggests it may be easier for females to support the young baby in its natural position than males. A straight body provides nothing for the infant to mould around.

As mentioned earlier in the book, arms are designed to support babies and children on either side of our body for many reasons, one of those being the way the arm works in providing a seat or "ledge" for their bum. Female forearms tend to be narrower than male, and this could factor into why it's so prevalent to see the variation of the shoulder hug (facing sideways with one leg higher than the other) on male bodies – the wider forearm supports a larger surface. It could also be an indicator that males are more suited to passive carrying than females as the baby will "outgrow" the space later. The female forearm is suited to be a supporting surface for the bottom and thighs of a small baby. The fact it becomes much harder to fully support the weight of the baby with one forearm as they get heavier, coupled with the limited length of this space, indicates it's designed as a temporary seating point. It can still be used as they grow and when they're asleep, but we tend to use it in a different manner: for example, as more of a stopping point for a sleeping baby as we angle their body so the shoulder bears much of the weight.

As mentioned previously, it can be harder to provide comfortable support with the forearm in a chest-to-chest carry for some months with slightly older babies. This position tends to work better for newborns and smaller babies, while their bottoms are narrower, and then comes into its own when they are big enough for the carrying person to use both arms – one forearm supporting each thigh when they are strong enough to independently support their upper body.

Hips

Evolutionally, a biologically female person was designed to be the gestational parent and primary caregiver. They are designed to feed their babies from their breasts which automatically tethers the infant to them, as well as the fact they have been physically connected to this baby during pregnancy. It makes sense that the body of the person doing all of this would be designed to facilitate carrying, so it comes as no surprise that biologically female bodies have wider pelvises than male bodies. This, of course, is usually referred to when talking about how the hips evolved to be just large enough to accommodate a large head to enable the baby to be born vaginally without increasing the costs of locomotion[4] but it goes beyond this. Whilst babies and children can learn to cling well to the biologically male body, the shape of the female body makes carrying for longer periods easier. The female pelvis is lighter in weight, the bones are thinner and smoother, and the iliac crests are flatter and more flared. Of course, in active carrying the baby doesn't sit statically on the hip bone, but it creates a more desirable surface to cling *around*.

Back to the energy saving effects in locomotion. Wider pelves take longer strides in relation to height and flex and extend less.[5] This is yet another suggestion that female bodies are preadapted for carrying infants as the desirable design for clinging does not impede on energy expenditure. Their hips also produce more of a sway from side to side when walking, which moves the child's body backwards and forwards in hip carrying. Essentially, it's providing rapid periods of greater and lesser clinging rather than a more static cling.

So, we know that male pelvises are narrower, and their height and width mean they are shaped differently to the wider and flatter female pelvis. How does this impact on carrying? Well, narrower hips create much less of a curve between waist and hip, along with the less desirable shape to cling around. Also, with some body shapes, the narrowness of the hips may contribute to an undesirable clinging angle. Triangular torso shapes are narrowest at the pelvis and get wider the further up the body you go. Now, imaging trying to cling to this shaped body - it would be more difficult, especially for longer periods of time. To cling at the same sort of height as on a female body the baby/child would end up with a backwards lean from the upper body, and the legs would need to grip onto the narrow waist with no padding or "shelf" below them. The firmer clinging surface combined with a less than ideal shape can end up being something of a nightmare for the clinger and carrying person. People with these sorts of bodies who haven't worked with the baby to help them learn to cling to their body from an earlier age tend to have more trouble with encouraging side clinging.

Another difference is that female bodies tend to have a greater carrying angle compared to male bodies.[6] This is due to them generally having wider hips – the arms and hands are designed to easily clear the hips. Whilst this seems to be an obvious design for the body (who wants to keep bashing their hands/arms into their hips?) it can also be a great benefit for in-arms carrying. For female bodies in hip carrying it places the supporting arm at its natural angle with a young baby rather than having to move it wider. For male and female bodies, if the arm is hanging naturally beside the hip, then is bent at the elbow to 90 degrees and the forearm is slightly rotated, still at its natural carrying angle, we see a greater or lesser gap between arm and body. Add to this the tendency for a larger waist-to-hip ratio for females, it appears female bodies are designed to have a biological advantage in hip carrying, as the additional space at the waist accommodates for some of the additional mass of the baby's body. As the baby grows and the angle of support requires increases slightly, the body gradually adapts to the increasing carrying angle as it also does to the rising weight.

Muscle

Biologically male bodies also have a much bigger volume of muscle even when sedentary, due to higher testosterone levels, making a significant genetic difference in strength. We know from the previous chapter that larger muscle mass comes from Type II, fast twitch muscle fibre, which is responsible for shorter bursts of strength. This obviously brings up more thoughts about whether this is an indicator of them having a biological predisposition to passive carrying, seeing as they would be better capable of carrying heavier loads and it's something to ponder. The issue with Type II fibres is that they tire easily but the further away from maximum output the weight is, the longer the muscle can endure it. So, for example, carrying a very heavy older child passively, whose weight is close to the most you can lift for a very short period of time, means you're not going to be able to carry them for long. Being able to lift say, 100kg, and carrying a 7kg baby passively is a different thing entirely. However, men also have a much larger quantity of Type I fibres in their arms too[7] which means they are better adapted to passively carrying the load which is their child.

Female bodies have less volume of the Type II fibre, just like babies and children. Both bodies are perfectly capable of building up both types of muscle, yet even if females regularly weight train by lifting heavy, they're incapable of building "bulky" muscles without having an extremely rare genetic makeup or using drugs such as steroids. The volume of muscle a person has can obviously interfere in the carrying process in some ways, by potentially altering both the available clinging space and the baby's access to it. Is it that far-fetched to consider the possibility of a perfectly designed system for the biologically expected male-female parenting unit? Would it not make sense to have different capacities which work in harmony with each other? The same with the baby or child – being genetically predisposed to build up endurance rather than short bursts of power aids carrying due to both enabling them to cling for longer and making sure their bodies aren't bulky. Of course, some babies are quite chubby, but fat is a different sort of bulk. When muscles are in use they're firm, whereas fat remains pliable whether we're active or passive.

It's interesting that active carrying relies on endurance. At first thought you may think that as a baby gets heavier you would need bigger, stronger muscles to cope, but the way active carrying develops for the clinger aids the caregiver and continues to be an endurance activity rather than a power task for the baby. When babies are lighter they're on a lower end of the clinging spectrum but as they grow and get heavier the clinging developmental process advances and they do more and more of the work. If the process is completed normally, the primary carrying caregiver's body adapts very easily to their load. Endurance is obviously still built up over this time, but the adaptations to the *clinging* load are different to the passive one. Passive loads feel as heavy as the amount they increase by. Active ones who are getting better and better at clinging don't feel as if they get heavier in the same way.

The most common form of holding of babies with good upper torso control that we see in the UK with men is an off-centre, high, one leg higher than the other carry which I nicknamed the "dad hold". This position seems to be a hip carry adaptation in my mind, as it's one-sided, enables the upper body to disconnect from the caregiver's to some extent, and there's asymmetry in the baby/child's legs. It facilitates communication with each other and their upper body is very much active in the hold. Of course, how the baby's legs are positioned makes the hip-health aware person wonder whether this position is great, and that's obviously something to be looked into going forwards, but it's something I think we'll probably find is OK for short periods of time.

Body fat

The average amount of body fat for a "healthy" female body is said to range between 21% to 33% of bodyweight and 8-20% for males aged 20-39 years.[8] This is obviously a wide range, and the actual *volume* of body fat that adds up to is going to differ depending on the person's weight. Having a greater fat percentage, and the fat being distributed in certain ways on the body, means that there tends to be more "padding" in all the right places for the clinger to meld into; fat is soft and makes for a pliable clinging surface. If you're going to be clinging to this person the most, you're likely to be carried by them for longer periods of time. If that is the case, doesn't it make sense for them to be beautifully designed to make clinging easier?

If we think about how the body gains weight over the course of the pregnancy, and holds onto fat in preparation for postnatal life, we know that most postpartum gestational parents aren't going to suddenly have little body fat. The NHS states that gestational parents gain on average 22-26lbs during pregnancy.[9] Some of this will obviously be the baby, amniotic fluid, increased blood volume, the uterus etc., but the rest is additional fat. Let's take the idea of a soft, pliable clinging surface and compare it to the clinging timeline. At the very least, it's said that it's 9 months on, 9 months off – it took 9 months to grow the baby and add the reserves, so it should take at least that to reverse the process. In the natural order of things, gestational parents have a greater fat percentage when their babies are younger, less developed and clinging less. As the weight loss happens (presuming a natural, slow decline in the additional fat) the baby is able to cling better as they're more developed. As I've said before, fat is squishy, and this makes for a surface which aids clinging, along with many other factors (such as skin, clothing etc.). When babies are less physically developed, nature has given them a way of transitioning to clinging, and the clinging surface changing gradually over time in relation to *their* development is ingenious. So, there you go, yet another reason (not that there should be any!) to throw society's expectations of the postnatal body out of the window.

As well as a slightly pliable surface being useful for aiding clinging, it also helps with stimulating reflexes. The drag against the skin helps with braking and firing off the reflex. The contact points are greater with a more moveable and flexible surface, giving the feet and legs something to meld with easier than firm, muscular surfaces. Male bodies are designed to have a much lower body fat ratio than female bodies, making them firmer and a less hospitable environment for developing clingers. As I mentioned, this doesn't mean that they cannot cling to these bodies, because they absolutely can, it just means that they appear to have not been designed to do the majority of active carrying. Leaner bodies require more work from the clinger, and more so from a body different to the one they have adapted to cling well to. Of course, not all will be lean, but it's a reminder that we're coming at this from an evolutionary expectation for caregivers to be at a healthy weight/proportion.

Postpartum bodies, lactating vs. non-lactating and pregnancy

Coming back to the postpartum body: it is shaped in a specific way. During pregnancy the growing baby and womb stretch the abdomen out and the body stores extra fat in expectation of breastfeeding. It takes around 6 weeks for the uterus to return to pre-pregnancy size and the body is designed to take many months to return to "normal". We've taken a look at the role of body fat and shape of abdomen in later clinging so let's now focus on the first couple of months postpartum.

Although society expects differently these days, the gestational parent is designed in a way which encourages them to slow down, rest and recover after birthing a baby. The tiring easily, aches after doing too much and copious amounts of bleeding are just some of these built-in "alarms" to warn them to slow down and recover. The size and shape of the postpartum caregiver is interesting. A chest-to-chest position is common for cuddling, as well as in early carrying. A natural instinct, it seems, to keep baby warm (lactating breasts have a higher degree of thermoregulation[10]) as well as enabling them to hear the familiar sound of the caregiver's heartbeat. In this lower on-body position the baby is in a prime place to stimulate their reflexes. The still-rounded, pregnancy-shaped abdomen provides a protruding part of the body for their feet and legs to easily make contact with. Of course, they can do so in other positions, but this is a very handy design which makes contact inevitable.

Post-birth, the caregiver is likely to be reclined when the baby is on their chest. This affords them a better view of the infant, relieves abdominal pressure from the still-heavy womb and means the abdomen is less flattened than when laying completely horizontally. The forces of gravity are exerted on the baby in a different way when on the diagonal than when they're laying horizontally. There may be a slight drag downwards on the caregiver's body and the abdomen is readily available. As we know from the chapter on breastfeeding, knowledge of postpartum traditions around the world[11] show us that stimulation of the abdominal region by others is popular and has benefits for the caregiver.

Although not all biologically female people will breastfeed their babies, from an evolutionary point of view the baby is expected to do so from this person. Male bodies were not designed to lactate for a baby, even though they can produce small quantities of milk. A caregiver's non-lactating breasts are smaller than their lactating ones, therefore the place the baby is shaped around is different based on whether or not the caregiver is breastfeeding. We know of course that they provide an advantage for preserving the curved spinal shape of the newborn and the older baby at rest, but what about hip carrying? Well, remember the upper-body disconnect which happens in active hip carrying? It looks like this cancels out any potential issue of breasts being in the way for many people.

Although it's possible to fall pregnant soon after giving birth, and many people decide to grow their families with smaller age gaps in the modern day, we know that in evolutionary terms this is a disadvantage.[12] A pregnant body is less able to carry actively and the additional load in passive carrying is not desirable for someone with already increased energy demands. The bulk of the growing bump creates a definite obstacle to hip carrying. Even though it's possible to lift the child's leg over the bump it's going to mean one of two things – needing to try and cling higher up the caregiver's body or using the belly as a resting point. Using it as a seat/ledge is going to create downward pressure which is not good for pelvic organs. Clinging to the back will also be difficult as the pregnancy progresses.

Multiple caregivers

Another little spanner in the works is adapting to multiple people carrying you. It may seem that it should be as simple as learning how to cling, then transferring that skill across the board, but if you think about it, every single *body* is different, and each person has their own carrying "style". If we were going to climb two different trees, one a tree we climb a lot and the other a new one, for example, we wouldn't simply move our body in identical ways to scale them. We would navigate the new tree differently for several reasons. One, they are different – the surfaces feel different, they have certain bits which help you and other bits which make it trickier. Two, the familiar tree requires less thought, less presence to what is going on – you've navigated it so much that it's second nature to interact with it and know what to expect back.

With humans, we have the differences in biology, plus the fact that within those categories we come in all shapes and sizes. It's easier for babies to cling to a slightly wider, malleable surface. When teaching a baby to cling (or attempting to overcome a side preference) it can be useful to bring them a little closer to the caregiver's midline when supporting on the hip, to start from a slightly easier place. This can be used with unfamiliar caregiver's too. There does come a point, though where a wider position is too wide to facilitate clinging in the younger baby and child. The carrying "style" of the caregiver will impact on the clinger's behaviour because they tend to default to how each person normally carries them. They soon learn what to expect in terms of support and are usually happy go with the option which requires minimal work if it's offered to them. If they know the caregiver does not provide passive holding in certain positions they tend to naturally go into the active stance as they're brought to or placed on-body. With a little knowledge, awareness to how they behave on different people's bodies, and some patience, new carrying people can learn how to best support the baby. The child can also become familiar with this new clinging surface and carrying can become more enjoyable.

Chapter 3.3

Side Preferences

Handedness

Another fascinating subject linked to physical development and carrying is the subject of handedness. It's been estimated that species-wide preference for the right hand was first seen around 600,000 years ago in the *Homo heidelbergensis* species.[1] These days it is estimated that somewhere between 70-95% of the world's population is right-handed.[2] It's a strange thing, handedness, and it's understandable to question why we have a dominant side of the body. What benefits are gained from having a side preference? There are different theories surrounding the subject, and one of those is that it is more efficient to divide labour up in this way.[3]

If we think about how there's a majority preference for right-handedness, wouldn't it make sense to see a right-side preference in hip carrying? If the dominant hand/arm is used in carrying it would surely make it easier to do, as it is better adapted physically to carrying a load? As straightforward as that may seem, studies have shown that there is actually a *left* side preference in carrying.[4] For right-handed people, an obvious answer for this would be that if you're right-handed then having that side occupied means you can't get on and do things with your dominant hand. However, researchers have found that side preference is usually independent of handedness.

There are two popular theories for this phenomenon. One is that in holding the baby in a shoulder hug the caregiver chooses the shoulder based on which way the baby turns to, to ensure they turn their head to look outwards rather than at the caregiver's neck.[5] In the study focused on this they found babies turned their heads to the right most, which is what caused the left side holding preference. The other theory is based on side preference in hip carrying and is based on a combination of brain hemisphere activation and interpreting emotions.[6] It's said that the left side of the caregiver's face most expressive, therefore the child can better read their emotions. This is coupled with right-side brain hemisphere activation in the adult, which picks up on the infant's emotional state.

Interestingly though, this left-side preference isn't replicated in all countries.[7] This brings me to the point that studies look into cradling and holding as opposed to active carrying. Some of these even use plastic dolls as the "baby"[8] which further complicates the issue. Carrying practices also vary around the world, with some cultures practicing predominantly passive carrying and others who use active carrying principles. I would love to see research conducted into side preference in active carrying. Obviously, we discourage sticking to one side, but the caregiver will more often than not have a preferred side at least; one which tends to get used a bit more than the other. There are some interesting things to investigate, such as are the caregiver-infant interaction when supported on each side of the body. Do they change in nature? What side is used most when walking longer distances? Does the baby have their own preferred side, and if so, does it match the caregiver's? If they are regularly held by 2 or more caregivers who only carry on one side (one being right, and one being left) does the child adapt to each caregiver, or is it harder to work with one of them? The questions go on and on. Another thing to look into is whether bias changes based on position. For example, we know saw the link to chosen side based on the newborn's preferred direction of head turning in the shoulder hug. Does preference change as the baby grows and transitions to the hip?

Caregiver preferences

So, we know for a fact that caregiver side-preference is a definite thing in holding babies. What we don't know is how much our own handedness impacts on our babies' own preferences and how carrying may impact that. In my limited personal experience, I found a definite, strong, right side preference from Isaac from around 18 months of age, as I have previously mentioned. It was strange to me, as he had experienced a varied carrying life before then, and it coincided with him stopping feeding from the left side. Initially, I chalked it up to being linked to the feeding preference, but I got to questioning it when I realised that we as caregivers seem to generate side preferences in our babies anyway. I know that I definitely preferred my right side even though I swapped sides more than most do and began to wonder how much influence I may have had over this.

In the following 12 months I worked on this preference to different extents. In the earlier months it kind of crept up on me, it felt comfortable – obviously, with my own preferences – and kind of appeared in full swing before I had a chance to see what was happening. Of course, it just felt that way because I was ignoring the signs. I guess from around 20-21 months it hit home and I started my attempts to reverse the preference. By that time, it was a firmly developed habit and Isaac was not going to take my attempts to get him out of this lying down. We made some noticeable progress but then the weather began cooling down and more layers were being added between us, and it felt like we hit a wall. Cold-weather carrying is very different to warm-weather! And so, we went through the autumn and winter with most of our outside carrying needing to play to his strengths, and using the warmer, indoor carrying to improve his left side.

I persevered over the colder weather, along with the unique carrying obstacles that in itself imposes on the process, and we came through the other side better than we'd left the summer. However, his preferences were still very much apparent and noticeable. Over spring we improved his strength and tone further, but by this age he was more aware, present and vocal about favouring the "easier" side. With the warmer weather I soon noticed a marked improvement in his strength on the left side, as he was able to work with less sensory barriers and get used to purposeful movement on the left side.

Babies' handedness and side preferences

Babies and children tend to develop a side preference in carrying and I believe this is at a bigger risk of happening in active carrying when the caregiver has a preference of their own and sticks to that side. It's still a definite possibility if they're only passively carried, and that may be to do with getting used to being in a specific position/place all the time. Simply put, they may be "conditioned" into preference. In active carrying though, if one side only is used then the muscles develop unevenly and eventually it will become difficult for the baby to cling to the other side if the caregiver decides to start evening out their carrying. I'm interested in the impact on other motor development and the child's movement in general when only one side is used to support clinging. Would this impact on durations they're able to walk before tiring, or affect the way they climb?

There's also the question of the impact on the baby's handedness going forwards. How much of handedness is impacted by environmental influence? I could find no studies focused on how carrying may impact handedness, but we can dissect the concept fairly easily ourselves. In a 2013 study[9] it was found that 39% of babies in the 6-14 months group showed a hand preference, with the majority being the right side. This increased to 97% having a hand preference for 18-24 months old, again the majority being for the right. If the caregiver is more likely to carry on their left side, it may seem like the baby would have a greater chance of developing left-handedness. It may not be that straightforward. You see, though a left preference does indeed improve the left *leg* muscles over the right, it doesn't favour the left hand/arm. When being carried on the left, the *right* arm has the dominant role in stabilising the upper body.

One way of exploring this further would be to observe the developed handedness in babies who are regularly required to cling in carrying, where the caregiver mainly or exclusively uses their dominant arm to support them.

An impromptu survey

As I was finishing up this book I found myself led towards another intriguing subject in relation to side preference. This time it was linked to breast refusal. I had noticed that each person (5 of us) who had commented on a post to do with breastfeeding had left-side breast refusers and I found it fascinating – both that for all of us the babies had rejected the left side, and that 5 people in a small geographical area had experienced it at all. It inspired me to post in a group of over 20,000 breastfeeders to see if this left-side refusal was just a coincidence. As I was writing the post I remembered my musings about Isaac's refusal of the left at 18 months, which coincided with his strong right-side carrying preference. I had wondered if there was any link between the two. So, I added a question about which side they would normally carry on, knowing most tend to just carry on the one side.

I had a much bigger response than I expected, and to begin with it showed that the breast-refusal was pretty much equal (no left bias) and that refusal seemed to be more common than any of us caregivers had thought. Whilst the tentative feeding bias theory appeared to not have much weight, there was an interesting pattern emerging. Although some only fed from the breast opposite to the preferred carrying side, many preferred the one on the carrying side. With this revelation I opened the question up to other breastfeeders outside of the group and sifted through the data. Unfortunately, many stated breast preference or refusal but did not answer the carrying question, and I didn't follow up with them to find out due to the enormity of the task. I also created a poll within the group so that I would get definitive answers. Of all the respondents I was able to gather 103 within 24hrs which answered both questions and showed a definite preference or (seemingly) permanent rejection of one breast. Upon revisiting the poll the next day, there was a massive increase in some of the answers, but I was unable to see all the names to cross-check for doubled up answers. Because of this I decided to do another collection, this one just of the poll results and answers outside of the group, and none from the first thread I asked on, to see if the results were similar. My unofficial findings are as follows:

First data gathering:

A

Left side refusal, right side carrying preference: 11

Right side refusal, left side carrying preference: 11

Left breast preference, left side carrying preference: 31

Right breast preference, right side carrying preference: 11

Total: 65 (63.1%)

B

Left side refusal, left side carrying preference: 9

Right side refusal, right side carrying preference: 1

Left breast preference, right side carrying preference: 13

Right breast preference, left side carrying preference: 15

Total: 38 (36.89%)

Second data gathering:

A

Left side refusal, right side carrying preference: 6

Right side refusal, left side carrying preference: 3

Left breast preference, left side carrying preference: 37

Right breast preference, right side carrying preference: 13

Total: 59 (64.44%)

B

Left side refusal, left side carrying preference: 4

Right side refusal, right side carrying preference: 2

Left breast preference, right side carrying preference: 12

Right breast preference, left side carrying preference: 14

Total: 32 (35.56%)

I grouped the four answers indicating matched side preference and the four with mismatched preference, naming them A and B, respectively. I omitted the responses which indicated a temporary side preference as I wanted to keep the question and results as disambiguated as possible. I also classed multiple children of the same caregiver with the same preferences as each other as one rather than counting individually. In cases where such children showed different preferences I counted these individually. The results in both the first and second data showed a majority preference of the breast on the side the child is carried on most.

One of the most interesting things this poll threw up was the fact that a whopping 30.1% of respondents in the first gathering had a left-side breast and carrying preference (LB/LC). In the second, this number rose dramatically to 40%.

This number may have been even higher had I been able to see all the names, as this was one of the figures which jumped exponentially overnight (19 additional responses). Most of the breast preferences came from the child rather than the caregiver. LB/LC also accounted for 47.69% of responses in the same-side preference group in the first gathering, and 62.06% of the second gathering. Added with the right-breast refusal and left-side carrying preference, the total comes to 64.62% of the first group and 67.24% in the second. This left-side carrying bias figure is close to the 66% seen in other studies. The left-side carrying bias was a little less in the opposing preference group, with left-side carrying seen at 63.16% in the first and 56.25% in the second. I also worked out the carrying side bias overall, independent of breast preference and found a 64.08% left side preference in first findings, and 63.33% in the second.

Although there appears to be a link between breast preference and carrying side preference, for this to be taken further it would be interesting to determine the reason for the breast bias. As the information gathering was unplanned in the first instance the survey was disorganised and didn't go deeper to create a clearer picture of *why* there were preferences and/or rejections. I would like to conduct a similar survey in a more organised manner in the future, which also obtains more information. Is it coincidental that it's more likely (in this sample, at least) for the preferred side to be the preferred carrying side? I would also like answers as to whether it's the side they are carried on that influences breast preference, as it appeared from the poll that carrying side-preference was determined by the caregiver rather than influenced by the child.

Interestingly, most people who mentioned handedness in their answers indicated that they carried their baby on a particular side so as to free up the dominant hand. In feeding, it appeared to be reversed – caregivers found it easier to support with the dominant arm. This conflicts with the general answers gathered though, as there appears to be a left-side bias with feeding preferences. However, this may be to do with baby's preferences impacting on the numbers. Many indicated the baby had the breast preference rather than them having a preferred side. With both there were a few opposing opinions, in that some carried with the dominant arm because it was easier and preferred feeding on the non-dominant side to free up the dominant arm. This would also be worth exploring further.

Additionally, it would be beneficial in the future to explore any potential links between active carrying and breast preference or refusal. I still wonder if one or the other impacted on Isaac's development of side preference and breast refusal. Although he is one child, so I cannot make presumptions based on an isolated experience, the timing of both appearing was interesting. I wish I had paid more attention at the time as I can't remember which developed first, or even if they appeared at different times. It wasn't until later on that I wondered about a connection between the two. I know that he would feed in-arms on the move so that in itself presents a possibility for a connection between the two. Maybe he found it easier to feed on the right side? Interestingly, he is very much a vertical feeder.

Active carrying and side preference

As we looked at towards the beginning of the chapter, the nature of clinging may lend itself to more diverse carrying. For a person practicing active carrying who is aware of the impact of sticking to one side, an awareness of discouraging side preference is present. It's reasonable to assume that – as in breastfeeding – the caregiver or child may have some side bias. However, where possible, we do not encourage a rejection of one side or preference of the other in breastfeeding. It doesn't tend to be comfortable for the person breastfeeding, especially if the baby is a frequent feeder, as the undesired side soon becomes too full when it's not being emptied. Whilst breasts are perfectly capable of taking on the role of providing the sole nutrition for one person, it tends to be believed that this is an adaptation so that it's possible to feed twins or different aged siblings without issue, or as a backup in case one side fails.[10] In other words, both sides are designed to be in use whether for one or more children, until one side stops working. We can take guidance from this for carrying. Although most people do have a dominant hand, carrying is designed to be a whole-body activity. It's not confined to one hand carrying out the task. We use both sides of our body during many different tasks and don't shun one side completely.

One of the most important factors in deciding whether we should take issue with side preference is the effect such bias takes on the body. We've already looked at what happens to the child's body, but what about the caregiver? In active carrying, good alignment is encouraged so as to minimise any potential negative impact of carrying through misalignment. This may look different between individuals due to each person's specific body and range of motion. If we only carry on one side, any slight or major deviations from normal alignment will impact the body every time we carry. If we alternate the sides we give each one a break, minimise negative effects and exercise both sides of the body. Even if a person were in perfect alignment, switching of sides enables a longer maximum duration of carrying as one side never need reach the point of fatigue. This goes for the clinger too. Alternating enables endurance to be built up on both sides of the body and makes it possible to rely less on outsourcing arms, if that is what the carrying person wishes.

Carrying happens often, every single day, over many months and even years. As such, it doesn't require carving any extra time out of the day so working on side preference can be as simple as using the non-preferable side as the first one upon picking them up. Other ways of minimising favouritism are using the dominant side when not carrying out tasks when carrying (in non-dominant side is preferred) or favouring the less-liked side when carrying skin-to-clothing or skin-to-skin.

As we found in the past two chapters, babies and caregivers' bodies are specifically designed in ways which support carrying. Neither possess these adaptations on just one side of the body, which suggests that both are meant to be used in supporting carrying as well as clinging. Added to this the fact the baby's muscles develop asymmetrically if one side is used exclusively or heavily favoured, it's clear to see that strong side preference likely should be discouraged where possible. There will always be at least a slight preference as its near-on impossible to divide the time on each side equally, just as there will usually be a side used slightly more in feeding unless they are strictly timed. Therefore, I conclude that a slight preference is likely a normal thing but should probably be discouraged from developing into exclusive or heavy use.

PART IV

Chapter 4.1

Evolution, Primates and Carrying

Studying newborn clinging behaviours has thrown up many questions for me surrounding what should be classified as "normal" for in-arms carrying. A discussion with my friend and colleague Ulrika about her thoughts on when and why babywearing may truly have been invented sparked deeper thought into the realities of on-body clinging today, and how different it may have looked further in the past.

For example, was babywearing invented because babies suddenly could not cling so well, due to the emergence of bipedalism, or was it something which has potentially contributed to weaker clinging tendencies due to the invention of such a "convenience"? Whilst I don't believe either of these statements are true for back then, it really makes you think about where we are today and throws up questions as to the long-term effects of inventions of convenience. We are not designed to be passive creatures. It's reasonable to speculate that slings may likely have just been used when babies were tired or sleeping, and in situations where both hands were needed, rather than as an ongoing replacement for arms. If a baby can cling well to you when alert, the need for a carrier is much less.

I want to briefly explore a part of the theory of evolution. Whether a person believes in it or not, there are certainly archaeological discoveries of interest. I'm not going to debate the subject – I would just like to take a look at some of what we know through discoveries, alongside some theories and see how they apply to carrying. There are many theories focused on the evolution of human nakedness[1] and the emergence of bipedalism[2], and all seem to have critical flaws. Nobody can agree on a definitive answer for the loss of hair density nor the reason for going from all fours to two feet.

We'll take a quick look at some of these theories to set the scene for my own observations, findings and theories.

The emergence of bipedalism would not have been instant. Bodies must gradually adapt over millennia to lose or gain features.[3] We also know that bipedalism is seen in some instances in species such as chimpanzees, our closest animal relative, usually when they need to carry a valuable possession. This can include the newborn infant as it develops its grasping abilities over the first month or two of life, and the mother will also at times use the other arm to aid walking.[4] We also know that quadrupedalism is possible for humans too. We develop into quadrupeds before we walk, for example. We can also, as older children and adults, revert back to all fours if we want to, however it is harder to sustain for longer periods of time due to energetic costs[5] just as bipedalism for non-human primates is more energy-expensive. The fact that humans are able to crawl, and some quadrupeds are able to walk shows that a moveable spine can be positioned in different ways to its dominant use to create other ways of moving.

For different species a dominant way of moving is formed, and for humans this is walking upright. The emergence of bipedalism with the retention of quadruped abilities, along with the fact that apes show bipedal abilities, presents an idea that it may have been practiced already and there was a key factor which encouraged a move towards bipedalism. The crucial point here is that it takes time to become adapted to a way of holding the body. This adaptation would need to be so widespread and taught over so many generations for a genetic adaptation to occur. A classic modern-day example of how long evolution takes is how America and Australia have been colonised for over 400 and 200 years, respectively. We have seen no genetic adaptation to the skin of descendants of the first colonisers of Australia nor those who took over the hotter parts of America.

If we look to the most recent archaeological discoveries we find some interesting information which may hold a key to the understanding of *how* bipedalism evolved. A near-complete *Australopithecus afarensis* (*A. afarensis*) skeleton of a 2.5-year old female, who is around 200,000 years older than the famous Lucy, was found in Ethiopia. The bone structure of the foot showed that bipedalism was well established, yet clinging behaviours were apparent.[6] The thumb-sized toe was curved and angled, which would be useful for grasping and clinging. This discovery showed that even if clinging and climbing were used much less as adults (which the advanced nature of bipedalism indicates), they were very much used in childhood. The structure of the foot was not a remnant of quadrupedalism as some have hypothesised. This revelation gives weight to the idea that clinging matters in this day and age, even if we use these behaviours less as adults. It shows that active clinging and bipedalism are compatible. Just as the mobility and abduction of the big toe was increased in the *A. afarensis* child compared to adults of this time, the mobility of the modern-day baby and child's foot and ankle is increased compared to adults'.[7] Our bodies are designed for certain movements and some of those – such as on-body clinging – are exclusive to childhood.

Another discovery – the partial remains of an *Ardipithecus ramidus*, nicknamed "Ardi" – uncovered evidence of even earlier bipedalism, some 4.4 mya. Study of the pelvis of Ardi revealed the ability to walk efficiently in a human-like way, whilst having the pelvic function for ape-like climbing.[8] The structure is different to other hominids and modern-day apes in that the upper pelvis is positioned behind the lower, enabling straight-legged gait. The lower pelvis of *A. afarensis* is closer to human construction, therefore it appears a loss of climbing power occurred over the 1.1 millennia between Ardi and the oldest *A. afarensis*.

Something I've found interesting in the comparison of non-human primates such as gorillas and humans is the difference in leg proportions to the rest of the body. Of course, we know that quadruped primates have longer arms than legs to make knuckle-walking possible, but there's a curious benefit to the proportions and shape of the legs when it comes to clinging. Human infants also have this proportional difference in their legs in comparison to older children and adults. Gorillas are a clinging species, both as adults and children. Human young are designed to be true clingers in infancy, but the nature of bipedalism means they will naturally not need these abilities to the same extent as they grow older. As we found in chapter 3.2, the curvature of the tibia – seen in other clinging primates – also straightens out. Bowed legs are suited for clinging, which is why apes retain this feature into adulthood. This seems to be indicative that human clinging behaviours change as we age rather than having a need to keep it in its entirety. Like *A. afarensis*, humans do not cling to trees or other objects in adulthood yet retain a need to climb. We're left with a changed clinging capacity, designed to work with specific environments and we retain the climbing capabilities of our youth. The terrain older children and adults are designed to navigate is not the same as apes. We're still expected to climb and use our bodies in ways which provide a wide range of joint motion and movements, it's just that an increase in sedentariness and man-made environments mean we're not given the opportunities we need for continued normal motor development as we age. It's abundantly clear that human babies and young children are specifically designed to be active clinging young and looking to evolution and primates to see the features which make clinging easier provides compelling evidence of this. It is a phase in their life which they will outgrow to a certain extent but is an integral part of their normal development.

The "loss" of human hair

A common misconception about humans is that we have evolved to be "hairless" when in fact we have just as many hair follicles as other mammals, and around the same density expected of an ape of human size.[9] The difference is that these follicles are much smaller. In fact, the average amount of hair follicles present in an adult are around 5,000,000.[10] Whilst this is obviously well-known in scientific communities, the term "hairless" is still used in general, and this fuels the popular idea that the hair which is left has little use to us these days. The main use cited for vellus hair (the tiny translucent hairs which cover our bodies) is for thermoregulation, and we're going to explore this fascinating body covering in the next chapter. The idea that humans lost their fur seems to have impacted heavily on the understanding of humans having still retained their clinging capabilities – even if they evolved during that time.

One idea is that humans lost hair density due to moving from rainforests to savannahs. An issue with this theory is that people indigenous to very cold countries have not retained or regained density to keep warmer. Also, it's been established that – since it's believed all humans originated from Africa[11] – that skin colour changed to being lighter to adapt to northern climates rather than us regaining hair upon migration.[12]

Another theory is that it happened because we learned to make clothes[13] so didn't need fur anymore, but current discoveries of clothing show that the oldest sewn clothing is 20,000 years old[14] and the oldest hide scrapings are 300,000 years old.[15] Furthermore, using genetic analysis of lice, it's been estimated that clothing may have first been invented 540,000 years ago.[16] It's estimated that humans have been "hairless" for 1.2 million years.[17] It is worth nothing that materials disintegrate, so findings may be limited due to this, but the analysis of lice does provide interesting information about the potential beginnings of clothing. A big question here is why did we not evolve to have thicker, hardier skin instead? This leads me to believe that the fact we have retained a covering of hair even though we have used clothing for so long (which keeps us warm), has some relation to carrying our young.

A new concept

Let me introduce you to the idea that we did not lose our ability to cling effectively. Whether you believe in evolution or not, the fact of the matter is that our skin and bodies, right here, right now, are perfectly designed to facilitate clinging behaviour. Whether or not bipedalism came from a new design (creation) or an adaptation (evolution) it's clear that there are specific adaptations and behavioural advantages for the clinging young who belong to the human bipedal class. Yet we're told over and over again that humans "lost" the ability to cling as they lost body hair, and that some clinging reflexes are vestigial. There is an overwhelming view that other primates are exceptional at clinging, and it's something we automatically associate with them. For example, we even have the term "cling like a monkey". How about we turn this concept on its head and reframe the way we define clinging. Humans are, in fact, made to be exceptional clingers, and possess many biological advantages which make this possible on an upright body - it's just that we cling in a different way to other primates, and this is what we need to accept. We need to let go of the idea that "good" clinging is defined by holding onto fur.

We're now going to take a brief look at how clinging works and is affected in other primates, purely to highlight how different ours and their clinging is. While I don't think it's particularly useful to constantly compare humans to other species when trying to understand our own – especially when it encourages wrong beliefs – there are some similarities and interesting findings which can help us understand how clinging works and what helps or hinders it. Beyond that, how a species adapts is species-specific. Just because we share 98% of our DNA profile with chimpanzees doesn't mean to say we are 98% similar. From this line of thought I'm going to look at how clinging depends on different factors in humans and apes.

The hair length, thickness, roughness and tensile strength varies between species, and this is directly related to the body mass of the different primates.[18] When relying on clinging by holding onto hair its density is crucial. The more hairs being grasped by each hand or foot, the greater weight can be withstood. Obviously, as the weight of the baby increases, so does the growth of hands and feet, meaning they're able to grasp a larger amount of hairs as they get heavier. Also, the length of hair factors into how many hairs they can physically grasp, so a clear pattern emerges between the properties of hair differing between species of different sizes.

Yet it's not just the weight to hair properties ratio, as how they sit on the body also comes into play. Very young chimpanzees cling independently in the ventro-ventral position from 1-2 months of age until their weight provokes a move to riding in the dorsal position around 6 months of age. At the beginning of back clinging, they are reported to lay high on the mother's neck and their bodies stretched out widely, meaning they are better supported by the mother's body. It's in older ones, who sometimes ride in the "jockey" position, that clinging with the legs becomes more important than before, as they sit more upright. The laying down position is still seen in older apes too. Things to note are that all limbs tend to be used, as well as the hands and feet, and the angle of the mother's body aids in their ability to hold on.

Interestingly, adult gorillas are much heavier than adult humans, but newborn gorillas are half the weight of human babies. As the frictional properties of the hair of furred primates decrease as the angle of the hair-on-hair load-bearing increases, it makes sense that these babies would need to be lighter – especially when riding in the ventro-ventral position – so that carrying can occur safely. The same goes for the fact that their hair can also only weight-bear to a certain limit. Therefore, it seems clear that the weight of human babies is not suited to clinging aided by the grasping of body hair.

I believe we, as humans, are extraordinary clingers. When you think about all the difficulties that bipedalism imposes on the clinger, having the ability to cling to an upright, moving caregiver – with or without long body hair to hold onto – is a tremendous feat. Human beings, as we have discovered so far, are shaped and built up in a way which is very much suited to in-arms carrying and upright clinging. We're now going to explore how our skin and body hair assists the process, providing yet more insight into the incredible science of carrying. This next chapter focuses on how clinging is aided in this day and age, but I hope the information from this chapter also encourages you to think more about how we've come to this point of having mostly lost the practice of active carrying even though we're so wonderfully designed for it.

Chapter 4.2

Frictional Properties of Skin and Hair

Now that we've explored some of the evolutionary theories and discussed some ways in which it can be useful or unhelpful to look to other species to try and answer questions about our different clinging capabilities, it's time to come back to the present day and explore the role of our skin and hair in carrying.

Skin

Skin is, of course, the largest organ of the body. Whilst it has many functions, it's role in clinging has not really been explored thus far. From an evolutionary point of view, the biological norm for carrying is skin-to-skin. Times have changed immensely, though, in the past half-millennia and most modern-day carrying in the UK (and, indeed, many parts of the world) is practiced clothing-to-clothing, or skin-to-clothing. Some of us are unable or unwilling to carry skin-to-skin, either occasionally or at all, but we need to look at this foundation of the carrying environment so that we can learn to adapt to carrying with barriers between us. It also gives us a clear understanding of other ways in which humans are designed as clinging young. Skin has frictional properties which make carrying skin-to-skin the easiest from a biological perspective. There are also no barriers to sensory stimulation, and a physical reaction of skin-on-skin – especially with movement added in – creates a sort of "sticking" reaction.

A quick jump back to species comparisons: human skin has much higher elasticity compared to apes. They need this less-movable skin to ensure holding onto hair to cling is possible. The fact we didn't evolve to have thick hides once we achieved bipedalism makes for a compelling argument that both skin and hair are connected to the survival of the species in a way we've not considered before. The way skin behaves in the carrying process of humans is very different to that of apes. In apes, the skin is a secondary reactor to the carrying process, whereas in humans it's the joint-primary one (along with vellus hair). Adhesion of person to person is not the only factor in creating friction. We need the pliability to meld with each other's skin and adapt structurally when stationary and moving to make clinging possible and to aid comfort.

To create this friction, we need either a load placed against our skin (in this instance, the baby or child), or a load plus movement. The two surfaces compress against each other and the load itself creates friction as it effectively drags against our skin due to its weight.[1] Kinetic friction then occurs when we start moving, which in active carrying tends to be walking. This increases the friction experienced and creates thermal energy. As with all materials there comes a point when the weight of the load becomes too much to sustain the friction needed to stop slippage. In carrying this means that clinging capacity must follow a positive trend in relation to the baby's weight. The heavier the child, the greater the need for the ability to cling to make active carrying possible, especially as active clinging relies on a point of gentle support rather than a pulling in of the baby's body. This also supports the unofficial observations that clinging increases when some movement is added in. The conditions of the skin change when kinetic friction occurs, with thermal energy stimulating sweat which we will explore further in this chapter. In carrying a high coefficient (μ) of friction[2] is desirable to assist the carrying process, making it easier for both in the carrying dyad.

Clinging relies on a certain amount of elasticity of the skin as the baby and caregiver must be able to connect to each other's bodies. As we found out earlier, fat helps with clinging as it provides cushioning properties below the skin's surface, which helps the bodies to melds together. Skin elasticity helps by aiding the movement of the combined surface. The ability for skin to return to its original state is also important, so that repeated carrying over many months and years doesn't permanently deform the skin, making carrying harder. It's interesting that skin loses elasticity as we age[3] and that the prime age for fertility in females is during the 20's. This may suggest a possible connection between the two in terms of biological advantages for the regular carrying person being at an age of reproductive prime. The fact that skin also loses elasticity with rapid weight loss is also interesting. From a biological point of view this also suggests it would be a hinderance for the gestational parent to lose pregnancy and postpartum reserves too fast. A slower use of these would be beneficial for both carrying and breastfeeding (in terms of sustaining the mother's body rather than aiding milk production).[4]

Skin's frictional properties vary based on things like the roughness of the surface, the weight of the load applied to it, the direction the load is bearing on the skin and whether it is dry, damp or wet.[5] Skin may seem fairly smooth, but it isn't. The top layer of the epidermis – the stratum corneum - is dead cells. These cells are keratinised (harder) and help form a protective barrier from the next layer. Whilst it has been found that removing the stratum corneum (SC) layer improves adhesion[6] it's also been discovered that something happens to the SC to increase its frictional properties when wet.[7] Therefore it appears the role of moisture aids the skin's friction when the SC layer is present. Moist skin is suppler and has a greater contact area. Dampness increases frictional properties.

Although we've discussed the importance of texture (and will explore this further), it's been found that skin's frictional properties increase when in contact with a smooth surface as opposed to a rough one.[8] This is because there are more contact points and a greater contact area is presented. This may seem confusing at first when thinking about skin-on-skin, but babies and young children tend to have significantly softer skin than adults. It seems reasonable to consider that the difference in smoothness may help with the friction between the skin of caregiver and child.

To throw a spanner in the works, the abdomen has been found to have the lowest coefficient of friction.[9] You would think it would have greater levels, being a contact point in clinging. But, the inner thigh has greater frictional properties. The same with the forearm versus the upper and lower back – the back has higher friction and the forearm lower. These opposing levels seem to work together to create a great "sticking" effect and is clearly observed and felt during carrying. There seems to be a trend with opposites in relation to skin and adhesiveness.

It would be natural at first thought to assume the areas of skin used in clinging and supporting the child should have the greatest frictional properties. However, this would have a negative effect on the developmental side of the clinging process. Humans are not designed to be passive. If the skin was Velcro-like, then the baby wouldn't need to participate much at all and there would be less movement and exercise in clinging. Measurements aside, simply carrying a baby skin-to-skin provides the caregiver with the insight that there is enough friction involved to easily support clinging, but not enough to comfortably provide independent clinging on the side of the body. From a bodily point of view, it also makes sense due to the fact the caregiver's arm would be displaced and hard to use for other activities when the baby is on-body in this position.

It's not surprising to me that active carrying is seen in cultures where women don't wear clothes on their top half and babies are naked or just their bottom covered. Maybe this lack of barrier has contributed to not losing the sense of how babies cling.

Moisturisers and other lotions

So, what happens when we interfere with the skin's natural state? Lotions can increase the frictional properties of skin, but if that area gets sweaty or is over-hydrated it has the opposite effect. Combining cream with sweat makes for a very slippery surface. This brings us to the modern-day invention of sun cream. Protecting skin from the sun is a big deal these days, especially with the added concerns of skin cancer. Many people choose to use it to protect their skin from the sun in this way, so it's worth us taking a quick look at this subject. An obvious way around increased slipperiness is making sure that any parts of the body involved in clinging which have cream on them have contact with clothing in the sort of heat where sweating-off of recently applied cream may occur. Clothing made of materials such as cotton can absorb excess moisture.

Before sun cream was invented, and in societies who live without this product, tend to practice safe sun exposure and others incorporate this alongside natural remedies which contain SPF, such as coconut oil.[10] This concept entails being aware of your skin's unique sensitivity and gradually building up exposure as the weather warms (in climates with markedly different seasonal UV exposure rates). The darker the skin, the more capable of filtering harmful UV rays[11] which is why – before more recent colonisation and immigration – people who lived in hotter climates had darker skin.

That most white skin darkens after suitable bouts of exposure to the sun shows that there is an evolutional adaptation to the seasons. The fact that skin (even dark skin) burns after too much exposure also shows that we are designed to spend time both in the sunshine and shade. It's worth taking a moment here to think about how this "development" of many countries has resulted in a drastic decrease in the natural environment. Trees, for example, provide plenty of natural shade, yet in countries such as England so much of it has been torn down and built on top of, creating unnatural living environments, therefore increasing our risk of overexposure to the sun. The more we mess with the natural order of things, the more we put ourselves at risk of harm. In summary, those of us with fairer skin are not designed to spend much time in the sun, so from an evolutionary point of view we *were* designed well, it's just that migration away from indigenous places as well as the stripping bare of the earth has meant we have to adapt to how these things affect carrying.

It's also interesting to look at how colonisation has affected mortality by increasing skin cancer rates exponentially for those with fair skin living in countries with a greater degree of UV exposure. For example it's estimated that a staggering 2 in 3 Australians will be diagnosed with some form of skin cancer by the time they are 70, and skin cancers account for 80% of all new cancer diagnoses.[12] This could suggest that our skin is pre-adapted to our ancestral native lands and that the optimal "carrying climate" also resides here. As we found out earlier, it's believed that skin adapted to northern climates as early humans migrated, and that skin got lighter due to a decrease in UV rays.

As a last point, interestingly, the volar forearm is one of the parts of our body most protected from sunlight. This means photo-aging is less likely and the forearm is protected from aging as much as other skin areas. There is much less deformation of taut skin, and there is a distinct lack of fat in this area of the body. The back also tends to have less fat. Whilst this may seem contradictory after I've already highlighted the benefits of body fat in clinging, in supporting the child it's less helpful. Clinging and supporting are different things. Clinging requires gripping, pressure, moulding. Supporting is aided by some degree of firmness. Too much deformation of the back and/or forearm would bring with it a need to increase pressure on the baby's back to achieve enough contact to create the friction needed.

Sweat and sebum

Our sweat can help to create more friction. We're not talking about sweating profusely, but the general damp skin effect. A bit like when you dry off after a shower/bath – your skin is wet at first but a quick "dry" may leave your skin damp. That slight dampness creates more friction than dry skin. In general carrying (with dry skin), creating movement, which is a big part of active carrying, the thermal energy generated triggers a sweat response from the body. This means that moving when clinging helps the carrying process. The fact that humidity increases friction is interesting. In hot weather it becomes harder to carry out physical activities, therefore it could be possible that this is a handy built-in mechanism to make clinging less taxing for the baby and easier on the carrying person.

There are differences between the coefficients of skin friction based on the location on the body.[13] I wasn't surprised to find a definite difference between the forearm and other parts of the arm and other areas of the body. The forearm is drier, meaning smaller quantities of sweat are produced. Babies and children also sweat less than adults, yet their backs – like adults – sweat more than other areas of the body. This information starts to build up a picture of how the way our bodies work complement carrying. It would be counter-productive if babies'/children's bodies were as sweaty as adults as it would make it harder to support them in carrying. It seems that we are designed to work together extraordinarily well, from the differences between adults and children's skin to the changes in how much we sweat from various parts of our bodies. There are a lot of adaptations of caregiver and child which again seem to work in a sort of "opposites attract" manner.

The adhesion of sweat to skin is what makes smaller beads of sweat stay on the skin rather than rolling off, and part of what increases the frictional properties of damp skin-on-skin. Also, the cohesion of sweat to sweat creates a binding of molecules which cause it to "cling" together.[14] The surface tension of the sweat also enables the sticking action. As we must add in the weight (or perceived weight) of the child in clinging, the right amount of skin moisture must be present so as not to be counterproductive. We also know that frictional properties are greater with lighter loads bearing on the skin, and friction reduces or is lost at a certain point with heavier loads. Again, this is why the clinging action evolves over time and the gripping strength appears to increase as they mature. There must be a delicate balance between weight, frictional properties and clinging abilities to counter the effects of gravity created by their increasing weight.

You may be wondering about when it is very hot, and we become very sweaty, and how this affects carrying. If the mutual sweating reaches a certain point, then it can be similar to the effects of lotions on the skin with the addition of sweat – it can create slippage. When we're carrying skin-to-skin, if it's so hot that the carrying surfaces are getting slippery, it's not going to be a temperature suited to carrying anyway, and it's not a common occurrence for this to happen during in-arms carrying. Excessive sweating is seen more in babywearing, where the addition of fabric traps the heat of two bodies increasing sweat production. It can be a very clever inbuilt "alarm" of sorts to stop, take baby out and cool down. We know that babies and children have a lesser ability to thermoregulate than adults so if babywearing is becoming uncomfortable because of a large amount of sweat then we're going to take notice. In contrast, this level of sweating is seen far less during active hip carrying due to the disconnect of the upper bodies. Also, when we're wearing clothes, sweat can enhance the frictional properties of the fabric surface in clinging too. Fabric can be a very movable surface. Clothes sticking to us provides less movement of the additional layer between each person's skin, which can aid clinging.

Skin development is not complete at birth. It goes through adaptations and maturation postnatally, many of these happening in the first 3 months of life.[15] Stratum corneum hydration levels stabilise at 3 months of age and skin roughness decreases. Sebum is produced by the sebaceous glands for skin lubrication and protection. Sebaceous glands form around 6 months post-conception and are responsible for producing vernix caseosa from this point to protect the skin from amniotic fluid. Post-birth, they produce sebum in high quantities, but this rapidly decreases in the first week until production is in small quantities in comparison to adults.[16] The role of sebum is to protect the skin from things which may penetrate it, so forms a barrier to the world around us, as well as preventing too much water getting in and out. The word "sebum" is Latin for fat or tallow. Its properties include wax esters and squalene, which waterproof the skin. Another useful property of it is its role in generating friction of both the caregiver and baby's skin. Sebum's role changes dependent on the temperature. In hot weather it emulsifies sweat which helps moisturise the skin and slow down the evaporation. and in cold weather it coats the skin to repel rain from our skin and hair.[17] Sebaceous glands are found in the dermis and merge with hair follicles so that sebum may be expelled through the follicle, coating the hair in the process.

Sebum levels are low in babies and children but rise rapidly at the onset of puberty. They stay at a plateau during adulthood before declining in older age. There is a major difference between males and females in that levels may fall sharply after menopause as opposed to a steady decline in male bodies.[18] The fact that babies, for the first week especially, and (to a lesser extent) for the next few months, are more "greasy" is quite interesting. It's around this time good head and neck control is achieved, and sebum production levelling off at this point in time seems to make some sense. Clinging is not happening before this time, and active carrying is supported by the hand at an appropriate level on the back to allow for movement. Sebum levels are higher on the back, forehead and chin compared to the rest of the body. Hands do not produce sebum so are dryer. This is yet another case of opposites aiding the carrying process.

Vellus hair

Our bodies are covered in tiny, translucent hairs called vellus hair. These hairs seem to aid carrying, but not in a "monkey" sense. If you've ever shaved your skin, or felt someone else's, you will probably have noticed how smooth it feels.

Smoothness equals slipperiness unless the opposing surface is rougher, as we found out earlier, as frictional forces are smaller on smooth surfaces. This is counterproductive in carrying, as we need the baby/child to "stick" to us, both through their clinging ability and providing a surface conducive to staying put. As we know, hairless skin isn't truly smooth because of the shedding of skin cells but shaving removes these, making it so. As we discussed, adults and babies/children have differing levels of skin smoothness, so it can be helpful to maintain this balance. Imagine caregivers and babies being completely hairless – how slippery they would be! Carrying suddenly seems way less appealing. With a lack of friction, we're not going to feel safe carrying our tiny babies, or older babies and children! It just wouldn't work.

The hairs help to provide a texture to the surface, creating help with creating friction. Providing the skin with more texture is comparable to how clothing with texture can help provide a "grippy" surface. The differences between adult and baby skin and hair again work together to provide more desirable clinging conditions. These wonderful little hairs are very much underestimated, as the main focus of their use is generally about how they insulate our bodies. While retaining heat is obviously a very useful thing, it wouldn't make sense if these hairs got in the way of carrying, would it? Just like so many other wonderful designs of our body, vellus hair has multiple purposes.

Hair follicles begin to appear at 8 weeks post-conception.[19] Lanugo hairs form in utero and start shedding from the head and face between month 7 and 8 and are replaced by terminal hair (scalp, eyebrows and eyelashes) or vellus hair. Babies and children are covered in vellus hair until puberty, when some of those will be replaced by terminal hair over time. Lanugo is very soft and silky. I find it interesting that vellus hair replaces it, and that lanugo starts to shed before birth; the "correct" timing being that little to none is left when a baby is born. Why might this be? Well, knowing what one use of lanugo is – it helps protect the baby's skin from amniotic fluid by helping the vernix casoea stick to the skin[20] – it makes sense that it would have outgrown its use once there was none left surrounding it. But, why does it shed before birth? Why doesn't it stick around after? After all, hair of all kinds has insulating properties and more hair would also mean more protection for the skin. Body hair was a positive trait to have in terms of natural selection, so we're told, so if it needed to disappear, why would it not be replaced with terminal hair?

Well, I propose that vellus hair plays a much bigger role in the survival of the species than just insulating our bodies to a certain extent. I believe this hair plays a huge part in biologically normal carrying and that it is a key component for skin-to-skin carrying. Some of this is to do with the sensory input we receive from hairs (which we will explore further in the next chapter) and some is to do with the frictional properties of hair. As we found out in the last section, sebum is expelled through hair follicles, therefore gives vellus hair frictional properties. Hair breakage is also less when wet than dry[21] and vellus hairs have disproportionately large sebaceous glands.[22] This appears to be a protective factor for the hairs as well as an aid for the generation of friction. Also, females have more actively secreting sebum follicles than males – another point worth considering when thinking about adaptations between sexes.

If we come back to evolutionary theories we come across one which states that bipedalism occurred due to the loss of human "fur".[23] In contrast, another agues the opposite.[24] Let's explore the latter theory, but from a carrying angle. Could it be a possibility that it was to compensate for the new angle babies had to be at when being carried/transported. This may actually have a lot of sense to it if we factor in all the wonderful frictional properties of skin and hair. To cling in the way that human babies do, long hair may be a hinderance. Whilst terminal hair also has frictional properties, long hair is more useful for holding onto in clinging. We already established that independent clinging on the side of the caregiver's body would not be useful as it wouldn't really free up their arm. If the baby is not laying on the carrying person to some degree, using them as a seat whilst holding hair, or grasping hair when hanging upside down (all seen in the way apes cling) the long hair loses its use in carrying. Therefore, I don't believe our lack of hair density is a disadvantageous trait.

Cold weather and goosebumps

The last feature of skin I would like to briefly discuss is "goosebumps". Goosebumps are thought to be a vestigial reflex, with the belief that its use would have been similar to apes in that it would make us appear larger to scare off predators when we had longer and thicker body hair. I respectfully disagree that this reflex is defunct. If we are cold we are not producing sweat, so how do we increase the frictional properties of our skin? Well, when we get cold the thermoreceptors in our skin send signals which make the muscles at the base of the hair (piloerectors) contract, causing the hair to stand up. This piloerection creates a bumpy surface, creating more skin texture. This may provide a temporary help for clinging whilst the bodies warm up with movement to create a thin film of sweat to improve friction. This reaction is also elicited when we are scared or in shock.

We also need to come back to the fact that the goosebump is created by the muscle in the *hair follicle*. This means that with every goosebump a hair is raised on end, and it helps make their locations more visible. For example, if you think you are naturally relatively hairless then you may be surprised to see goosebumps on your forearm, abdomen and so forth. What may the benefits be of raised hairs? Well, there is evidence to suggest that hairs interact with each other[25] (e.g. tangling and sticking to each other) and this reflexive movement may produce a more effective interaction between baby and caregiver's hairs.

Another effect of the cold is on muscular contraction. It's harder for muscles to contract at lower temperatures. Temperature affects the rate at which oxygen is released from haemoglobin. This means less oxygen is available to the muscles, making contraction more difficult. Now, at first thought this may seem like a hindrance to the carrying process, especially if we couple this with the fact that energy expenditure is increased by the body attempting to warm ourselves up, but we actually have a reaction which makes the carrying process easier when caregiver and/or baby feel cold. The resulting effect of limited oxygen to the muscles is a stiffness in them. This tension produces a stronger, reflexive, clinging effect.

Chapter 4.3

Senses in Carrying

As we discovered earlier in the book, motor development comes from the maturation of the central nervous system (CNS) through sensory stimulation. Sensory input triggers motor output. As the CNS matures, reflexive behaviour develops into more complex, voluntary actions. Although clinging becomes voluntary much of what is going on is happening in an automatic, subconscious way. For example, a child may lose some of their grip with their legs, which may trigger a more reflexive response of suddenly gripping tighter (a survival response) before the voluntary action of readjusting themselves occurs. Or their caregiver moves in an unexpected way and their arm automatically braces against the upper arm. In this chapter we will gain a better understanding of why babies' senses are at different stages of development at birth and how this on impacts carrying. We'll also explore the ways in which the body processes this information as they mature.

Perceptual systems

Perceptual systems are also involved in carrying. We're going to take a brief look at some of these before we explore other senses in carrying.

Proprioception

The baby's sense of relative position of their body parts and limb position obviously comes into play when they are participating in carrying. To learn how to cling voluntarily, they need to have an awareness of where their limbs are and what they are doing, as well as what is going on with the rest of their body. Before they reach this milestone, they have proprioceptive reflexes such as the upper- and lower-body placing ones we looked at in chapter 1.2. These make the body automatically reposition itself when needed.

The proprioceptive system receives information through stimulation of the muscles, joints, ligaments and bones. This is then processed and tells the child how to position themselves. If something isn't working right – for example, they find themselves not able to cling properly – this sense alerts them to where their limbs are and what they're doing. Then they're able to adjust their body back to their normal clinging position.

Equilibrioception

The child's sense of balance also factors into carrying, as this communicates to them when they need to adjust themselves and in young babies also triggers certain reflexes to correct position, as we saw earlier in the book. Eyes, ears and proprioception all work together with the vestibular system to signal when adjustments are needed.

Simply by moving we trigger this sense via the vestibular system and the feedback received by the baby/child lets their body know on a subconscious level what subtle actions they need to perform to maintain their balance on-body. On a conscious level they're directed to do things such as shift a leg or grab onto the caregiver's body to steady themselves. If we lean forwards, they will feel the change in balance and should cling tighter/grab on. Every time we take a step we're disturbing their equilibrium, so this sense is in constant use in active carrying.

Kinesthesioception

The sense of acceleration is helpful for sending a message to the baby/child to cling tighter. When practicing active carrying it's very easy to feel the change in clinging behaviour if you've been walking and suddenly need to break into a run. From an evolutionary and modern-day point of view this reaction makes sense. This is, at its core, a survival response. If you're running with a baby or child there logically would be good reason – it's not something we incorporate into daily life. In a life-threatening (or otherwise) situation, you need the child to have a reflexive response to the acceleration rather than be destabilised. The rapid motion's effects on friction are going to mean slippage, so the only solutions available are increasing our support or stronger clinging from the baby/child. The immediate response to the sudden movement tends to be an increased intensity of clinging, followed by a slight relaxation. Clinging hard – beyond the normal intensity of regular carrying – uses the fast-twitch muscles, so cannot be sustained for long periods. Basically, an intense response is invoked to counter the sudden movement, then after a short while clinging goes back to normal.

Another response tends to be increased clinging behaviours from the upper extremitiy/ies – especially in older babies and children – which also assists the baby and caregiver. They tend to turn in towards the carrying person's torso, which serves to protect the body and head from the impact of running. It may seem counterintuitive when we know that full-body contact tends to be associated with passive carrying, but the engagement of the upper body counteracts this. There also tends to be a reaction from the carrying person too. This can be firmer support as well as the free arm providing additional support to the head and neck.

Sight

It's interesting that babies' sight develops in ways in which seem to impact on carrying. To begin with, babies' vision is far less developed than an adult's.[1] The ability to focus on objects is limited, meaning they're ideally suited to gaze at their caregiver's face when feeding. Although vision is underdeveloped, it doesn't mean babies cannot process what they see. Studies have shown that very newborn babies (as young as 12 hours old) can recognise their mother's face when shown theirs and another face on a TV screen with no sound.[2] This is one of the many ways in which we can see how intelligent babies are and how very quick they are to learn, right from the get go.

There's not a massive amount of carrying going on at this point in time; instead, there's a big focus on holding, cuddling and resting together. It really makes so much sense for sight to be far less developed than other senses at birth. Can you imagine living in darkness all your life, then suddenly having the capacity to see very clearly and at great distance? We have a semblance of understanding of what this might be like if we're sleeping in darkness, wake up, and switch a bright light on. It's complete overload for the eyes, painful, disorients us, and takes a while to adjust to. Yet that's when we know our surroundings and have fully experienced the ability to see – how different would that be for us if we were experiencing full sight for the first time? Add to that the fact that the newborn has experienced only one environment its whole life, and in limited ways. For example, touch has been the primary sense in the womb, and that was limited to their body, umbilical cord, amniotic fluid and uterine walls. They've not learned about the environments outside the womb in this time and having limited sight, but a highly developed sense of touch means they're able to get acclimatised to the world in a way in which they're already used to.

All they need right now is to get to know the outside of the person they called home, adapt to a different way of being nourished and work out how they fit into this new world. They have no concept of being carried in-arms. Their only reference point is how they were carried in the womb, and that was different in many ways. This is why stretchy and woven fabric slings tend to be very popular with babies from birth – they mimic the firm, full-body hold of the uterine walls. It's familiar; especially if the person using the wrap is the person whose voice and heartbeat they know best.

So, it's clear that good sight isn't needed from birth, and it could potentially hinder the carrying process if it was well-developed at birth. Initially, the focus must be on learning the new carrying environment – the outside of the body – and underdeveloped sight facilitates this by keeping the baby's focus where it's needed. If they had full sight the distraction and visual overload would likely be overwhelming. There would be too much to take in and the focus would be away from their caregiver.

As their sight develops further over the next couple of months they'll be able to distinguish between colours, see further and their peripheral vision will improve. This leads to an interest in the things around the dyad in carrying and is very noticeable in the shoulder hug as they lock onto certain objects with their eyes and sometimes even communicate with what they see. Something I've always found fascinating is how they can appear to be looking at a blank spot (e.g. a cream coloured wall with nothing on it) and start having what sounds like a chat with it. Whether they're interacting with a plain or interesting (to us) object, this more often than not triggers communication from the caregiver, amused at this seemingly one-sided conversation. This can help to form the beginnings of 2-way communication in carrying, with the carrying person noticing the baby's visual interests and taking their cue as to where to take them to and what objects they like to look at. These early building blocks of communication between the dyad pave the way for more expressions as their language centres develop further.

With the growing newborn, having them in a shoulder hug (which is a very natural go-to hold for many caregivers) positions them right by the caregiver's face. This ensures close visual monitoring from the carrying person. As they rapidly grow and spend more time awake their sight is also improving, including emerging depth perception.[3] The shoulder appears to be a great place for them to spend much of their early active carrying time as it enables them to be at a place on-body where their vision is not restricted. As they develop further they're able to see more and let what they see guide how they move and interact in this position.

Around four to five months of age their binocular vision and spatial perception is further developed[4] and this usually coincides with good upper torso control. More refined vision helps them to better understand the world around them. At about 5 months old they're noticing smaller things and are able to examine objects closely, aided by the control of their learned grasping action. This interestingly links into transitioning to hip carrying. Being able to spot all sorts of things around them and being held in a position which makes this all the more possible is a harmonious combination. The further their sight improves, the more developed they are in other areas. By the time voluntary clinging is established they're able to enjoy the world in visually superior ways. By the time babies are a year old, they should be able to see as well as an adult[5] and body control plus clinging behaviour is well established by this point. It's fascinating how sight develops in a way that's appears to be linked to how well they can control their body to take in what they can see, and how they can interact with these things.

So, it's clear that development of sight also aids other physical development, by providing interesting visuals which encourage movement of the head and neck, then the body too. Focusing on points of interest encourages them to hold their head up and as steady as possible, which aids carrying by increasing their stability, moving them closer towards being able to cling to their caregiver/s. As time goes on and their sight improves further, it will play a big part in sending them information as to where they want to be positioned. For example, if their vision is being blocked by being chest-to-chest and they want to see more, the restriction triggers them to communicate they aren't happy being in that position.

As well as the practical function of sight in carrying, it also plays a role in emotional development.[6] Eye contact is a form of non-verbal communication and can have an important role in forming connections with people.[7] It's important in infancy as it forms a part of their understanding of how social interactions work. In the most-used form of active carrying – the hip carry – the baby is at a vantage point to be able to easily connect with the caregiver visually. In a hip carry facial expressions are easily picked up on and the vantage point offers better interpretations for the dyad. As babies get older they spend more and more time in an active state and each time they are afforded the chance to practice clinging is a wonderful opportunity to build on the baby-caregiver relationship through how humans interact with the world and each other through sight.

Smell

In the early days smell is one of the ways in which a newborn baby navigates their environment. This sense is more developed than some of the others, and this is likely due to the fact that it helps them to find the breast and gives them a way of reflexively recognising different people.[8] The smell of their primary caregiver is the one they recognise and seek out the most. The fact that carrying provides close proximity to that smell seems to be another one of the reasons why babies want and need to be held so much - this sense sends them messages about where they are. With their limited range of vision in the beginning, it's easy to understand why a familiar, comforting scent may be sought.

For the caregiver, carrying obviously puts the baby in close proximity to them. In the coming weeks, as we know, the shoulder hug is a regular carrying position. This puts the baby very close to the caregiver's face. The scent of the baby is easily detected, and with a slight turn of their head they're even able to touch them with their nose, gaining the most olfactory feedback possible (as well as tactile). The scent of the baby is, arguably, just as important to the caregiver as theirs is to the baby. It enables the caregiver to recognise them, and for many stimulates pleasurable feelings. This initial carrying position offers an environment suited to getting to know the baby in a way where it's not essential to connect with each other via sight. It effectively brings smell (and touch) to the forefront of sensory feedback. This position also promotes interaction and feedback away from distractions from other smells, which will be even more helpful if the person carrying them is lactating. In contrast, for a well-fed and settled baby, being held to the chest may be of comfort to them and may provide a place of calm and contentment. We see babies both in-arms and slings/carriers relaxing into this space and finding themselves drifting off to sleep or relaxing in the quiet-alert state.

As they get older they may rely on smell less in carrying, but it is likely to still be a big factor for them feeling safe and secure, especially when they lean into the body in passive carrying. There are also many suggestions that smell has a significant role in cognitive development, especially from feedback from mother-baby dyads[9] so the act of keeping them close in carrying – where they are better able to gain olfactory stimulation from the caregiver – may be a useful environment for this. It's also been shown that oxytocin can be released via the sense of smell[10] and we know that this hormone plays an important role in bonding. The continued benefits of this as they grow older are more likely gained in passive holds, where their body and head is closer to the caregiver.

Sound

Hearing is well developed at birth yet will still take around 2 year to mature completely.[11] Hearing is also useful for sending information to the baby about who they're with, therefore where they are. They've gotten to know the voice and heartbeat of the gestational parent whilst in the womb, and this is obviously the primary way of recognising them on the outside in the first instance. Being on-body – especially resting with their ear on the caregiver's chest – allows them to hear these familiar sounds in a comforting way.

Some sounds may also cause upset to them (and may trigger the startle reflex) and being in-arms, they are perfectly placed to receive immediate comfort. This new world without the muffling barrier of the womb can provide auditory overload too. As with all of the senses, it's so clear that babies are designed to be on-body much of the time in the early months as they get accustomed to the world on the outside.

Touch

The complexity of the ways in which touch works is astounding. There are so many ways in which this sense is used in carrying. It's probably the most obvious one as the physical act of carrying has a strong focus on holding, touching, supporting and clinging. Touch triggers many of the carrying reflexes and sends signals to the older baby, letting them know what action they need to make to keep clinging. It sends us information too, telling us when we need to adjust or change their position, or switch our support.

This sense is the first to develop in the womb[12] and is highly developed at birth. It is, of course, the main way in which babies interact with the world in the beginning. The receptors of the skin are linked to the spinal cord and touch – like the other senses – plays a role in developing the baby's nervous system. The fact that this sense is vital for normal development and that carrying requires lots of touch is interesting. As we've found so far, caregiver and child are built physically to facilitate carrying, and the addition of the way senses play into this is yet another way humans are exquisitely designed. We're now going to explore the importance of touch in carrying from two angles – practical and emotional.

Mechanoreceptors in the skin respond to the varying amounts of pressure applied from the support we give and have this amazing ability to communicate subconsciously to the child that they need to cling more, or less, depending on what area of the body they are stimulated at, regardless of the amount of pressure applied. For example, when support is moved from the upper to the lower back there tends to be an almost automatic change in clinging. A message is sent to the baby to relax their legs slightly. Hairs also function as sensory receptors; sensory nerves surround the hair follicles and react to pressure on the hair shaft.[13] This is another reason why vellus hair is so important.

The sensory receptors of the skin are:

- Free nerve ending (perceives pain, touch and temperature)

- Merkle's disc (responds to light pressure and perceive fine differences in location)

- Perifollicular (perceive when the hair on body/face is being touched)

- Ruffini corpuscle (responds to touch and pressure)

- Meissner's corpuscle (also perceive differences in location – located in palms and fingers)

- Pacinian corpuscle (sensitive to pressure and vibration)

Pacinian corpuscles help us process touch in clinging on a conscious level and are responsible for the caregiver feeling how the baby is sitting on them, and for the baby feeling how they're clinging. Ruffini corpuscles are sensitive to changes in position so play a role in proprioception.

The types of touch used in active carrying are primarily practical. This should not be taken negatively or thought of as being inferior to holding and cuddling due to the lack of focus on emotional benefits. No, active carrying provides vital benefits for development and sensory stimulation. It also has secondary benefits of connection with the carrying person, as well as providing opportunities for gentle touch too. The practical touch benefits can be useful for sending messages about what a person needs to do in the carrying process non-verbally. One study found that a certain type of touch (along with an audible gasp) may convey danger to the infant and influence them to proceed with caution.[14] Non-verbal communication is useful in carrying.

We explored in some detail in chapter 2.3 how physical communication occurs on a conscious level in carrying. This requires a voluntary action in response to the message received from tactile stimulation and is why it can be harder for caregivers to respond appropriately if they don't understand the meaning of certain pressure or movements from the child. For example, when babies use their legs and feet to communicate a want for the caregiver to change direction. However, if the message is understood by the carrying person (or clinger, with feedback such as slippage) it is possible for a more reflexive response to be developed. Other tactile feedback received on a subconscious level also helps the caregiver carry – and clinger to participate – "reflexively".

From the emotional point of view, it's again fascinating how big a role touch can play in communication. Another interesting study showed the power of touch at a level even the researchers were astounded by.[15] They wanted to see if it was possible to convey emotions to another person simply by touch. To do this they made sure the participants couldn't see each other, and they expected a positive response rate within the realms of guesswork in decoding the emotion being directed at them. They were blown away to find the actual rates as high as 78%! This has massive implications when it comes to carrying our babies and children. If we can convey something as deeply complex as emotion through touch, what more is going on when we're not even thinking about it? Also, the opportunities for intentional communicative touch are frequent in carrying, and this may be a way of building a deeper communication between baby and caregiver.

A big question needing answering is "what are the implications of not carrying skin-to-skin?". It's been shown that holding babies affects their DNA at a cellular level.[16] What changes may happen when there are no sensory barriers? We know that the skin receives many messages and that these are best received directly. We also know the many benefits of kangaroo care, even for healthy, term babies.[17] The different ways in which touch facilitates the dyad's communication are many, as well as the bonding benefits. It would be interesting to explore whether the non-verbal, tactile communication in carrying – both clothed and skin-to-skin – aids in the dyad's abilities to understand each other in general. This is especially worth investigating further as it's observed through certain practices (such as breastfeeding and elimination communication) that signals are more easily read on-body. If these are recognised well on-body it would be natural to postulate that this could aid caregivers to read those same cues better off body once understood.

I'm a great believer in the power of positive touch – especially skin-to-skin. In a society where it tends to be unacceptable to touch each other unless you're close (and even then, not that much), some of us are touch-starved. When this happens, it can sometimes make positive, platonic touch almost painful – causing a person to either feel discomfort or flinch when it happens, even if we think we're ok with that person touching us. Add to this the multi-layered issues that different people experience such as birth trauma, other past or present trauma, our own attachment to our parents, systemic racism, systemic sexism, mental health issues and more, it becomes clearer that as a society we're very much broken.

The fact of the matter is that there is so much power in the act of consensual physical affection. We see it so clearly when interacting with babies and children – positive touch can soothe worries and anxiety, calm them down, alleviate fear, reduce pain, convey love, release oxytocin, endorphins, serotonin and so forth.[18] It can even reverse the effects of prenatal stress.[19] With our babies and children, touch tends to come easily and happen frequently, as they need us to pick them up, feed them, clean them and care for them in other ways. As they grow older this lessens as they gain more and more independence, and it can sometimes get to a point where it's easy to forget to make time for this focused, present and meaningful touch. I believe it's important to keep reciprocal positive touch going for as long as possible – it's just as powerful in adulthood.[20]

We see the effects of a certain type of touch in babywearing and swaddling - the all-encompassing constant hold of the fabric around the baby. In babywearing this is a replacement for arms yet provides the closeness babies want and need, keeps them in constant contact with the caregiver's torso, and holds them in a way we can't replicate in quite the same way with arms. It also shows us that a firm hold is soothing, calm inducing. Creating this exact sort of pressure with arms can be difficult with little people and may be why so many people find that when nothing else works, a sling does.

Carrying can provide a constant, steady stream of positive touch, in a different way to babywearing, stroking and intermittent touching; even holding. Of course, we're entering new territory here. With countless studies conducted on different areas of human touch, none yet are looking at the touch of active carrying. I can only present my beliefs based on my observations, other research and the theories which have developed from these. One of my theories is based around the amount of pressure and how it's produced – that it plays a big part in how the effects of touch are processed. It seems fairly straightforward if you compare gentle, loving touch to rougher, playful or even aggressive touch. Even thinking about the sort of touch in passive carrying seems to conjure up gentleness, calm and contentedness. What comes up if we think about the touch involved in supporting clinging/active carrying? For some who carry passively, it may bring up ideas of a firmer touch more detached link to the carrying process, but this couldn't be further from the truth.

As we've learned, a clinging baby doesn't require their caregiver to pull their body into them – that would mean the caregiver is doing more of the work and would promote passivity. Witnessing active carrying can bring prejudiced ideas as to how the bodies interact. For example, take the classic active hip carry support points. For some people who I've done this active carrying exercise with who also babywear, they've held beliefs that the support provided could compromise the baby's back. When they actually feel what the arm is doing in relation to the baby's back and realise exactly how much the baby has to participate when you provide just a buffer, it begins to make more sense. What you think you see isn't always what is actually going on. Providing support on certain parts of the body sends messages for the baby/child to cling actively yet supporting other parts (such as the legs) tells them to relax and disengage from the carrying process. It becomes clear that touch in carrying is especially important when we remember that it plays a vital role in how babies learn. It's easy to recognise this when we think about how babies explore the world around them, learning through tactile stimulation. Yet touch from others provides learning experiences too, as I briefly touched on earlier.

Another thing about touch in active carrying is the potential link between touch and trust. This is apparent in the child who has sensory issues relating to balance and requires the caregiver to anticipate the effects of movement on their vestibular system. A firmer or different sort of touch may be required. For example, one may need to be hugged into the body (so a focus on passive or semi-passive holds) and another may be ok with the addition of the caregiver's free hand holding theirs or touching another part of their body (e.g. face, arm). The caregiver's body is meant to be a place of safety and comfort, as well as a dynamic learning environment. As well as there being an element of trust in active carrying – even for the baby without any pre-existing issues – there's the point that they also are experiencing new situations in-arms. Being able to navigate these experiences with a trusted caregiver only seeks to associate the caregiver's body as a safe place.

We know that babies thrive on a combination of touch and interaction and these come easily with active carrying. The baby/child's perceived weight being less means we're likely to be more present and readier to engage with them than when they are passive and feeling heavier, weighing us down. However, it's important to remember that babies can get touched out too - sensory overload comes in many forms and carrying can contribute to it. For a non-touched-out baby with other sensory overload, they're in the perfect position in a hip carry to lean in and disengage from the world, reconnecting with the carrying person through touch and contact with all of their body, including the baby's face to their chest. This can communicate to the caregiver to move them to a position such as the passive chest-to-chest or shoulder hug holds where they can more easily disengage and receive physical comfort.

As a final note, we know that skin-on-skin provides the most easily understood physical communication. Sensory input is stronger with no barriers and bodies tend to work almost intuitively together. This is one of the reasons why skin-to-skin carrying helps so much with caregivers understanding the carrying behaviours and how the baby/child fits to and works on their own body. The following chapter explores the (often inevitable) barriers to sensory input.

Chapter 4.4

Barriers to Sensory Input

There can be many barriers to sensory input, and not just physical ones. We're going to examine different barriers in this chapter and work out ways in which we can adapt to them to make carrying easier.

It's interesting to feel the difference between clinging when skin-to-skin, your skin bare and theirs clothed, yours clothed with theirs bare and both of you clothed (at different levels of thickness and types of fabric). The sensory input (less or more), the reactions of certain fabrics against skin, and against each other – the list goes on!

Anything preventing full skin-to-skin contact:

- Gets in the way of sensory input, for both baby and caregiver

- Changes the nature of gripping mechanisms

- Can interfere with the range of motion the baby/child has in their lower body (e.g. denim)

Of course, in the UK, most of the time one or both will be wearing clothes. It's simply impractical – especially in the colder months – for many to do lots of skin-to-skin carrying. As mentioned in the previous chapter, however, it's useful to do at least some skin-to-skin carrying to get an idea of the normal behaviours of the baby on the body and to understand better how the carrying dyad's bodies interact with each other. Not everyone will want or be able to do this, and that's where we can use common factors to create a more "universal" blueprint to work from – it just won't account for any unique carrying quirks and individual issues, but these can usually still be picked up on if seen in other carrying too.

Thankfully, although clinging is affected by sensory input, it's not completely inhibited by layers between bodies. In fact, it can work very well – just not as well as when there are no barriers. Being aware of the fact that *every single layer* and type of clothing changes carrying to a lesser or greater extent helps us to understand and modify carrying when clothes are wanted or needed – which tends to be the vast majority of the time. Let's break this down into sections, from one layer on one person to multiple layers on both people.

One layer on caregiver

So, the first barrier to sensory input from the skin is if we introduce one layer of clothing on the caregiver, as this is the first step away from skin-to-skin for the baby, the one whom we want to be doing more work, and who needs the sensory input more than us. Their legs and feet uncovered means that they still have the sensory stimulation from bare skin, but they are removed from the "natural" environment of the caregiver's skin. This means they are also working that much harder to hold onto the caregiver – how much so depending on the fabric type. The fabric creates a new environment from which to work with. Unfortunately, this is not a living one which can respond back, but thankfully the caregiver's skin is under this, so they are able to respond to the messages which filter through. They are able to help the baby based on the feedback they receive, maybe by lifting them up and placing them back on-body in a slightly different way or changing where they're providing support. Fabric is a more movable surface than skin. There is, of course, usually a stopping point but this all depends on the type of material in question. For example, a top with elastane will move more than a 100% cotton one. Stretchy fabric – especially if not fitted to the caregiver or child – doesn't always cope well in carrying. It's much harder for them to cling to fabric which has a noticeable amount of give to it, as the stretch pulls their leg down the caregiver's body. Even strong clingers can be affected greatly by this.

The baby or child is also able to learn to navigate this new "skin" by the sensory feedback received via their bare legs. An older baby or child, for example, will be able to experiment with where they can get better grip from, adjust their legs for better comfort, change how they're holding on with their upper body to use their lower extremities less, or signal to the caregiver that they need some help.

One layer on child

Next up is a layer of clothing on the child, with the caregiver's skin bare. In this instance, they have a block to sensory input, but the caregiver does not. If the material on the baby is not the slippery sort – the caregiver's skin will provide some grip still, just like when they're clothed, and the baby's legs are bare. They do, however, have to contend with dealing with the sensory input from the inside of their clothing. If it's not skin-tight and also of a slippery nature then it can give off all sorts of information, such as signalling that the surface they're trying to cling to is not stable.

We must also remember that the friction levels of fabrics against skin vary. For example, polyester and wool create a greater coefficient of friction on skin than cotton does.[1] Fabric can also give conflicting messages about what they can do with this extension of their own skin. For example, if the inside of their clothing has a rougher surface than the outside (e.g. non-fleece-lined joggers), the message they are immediately getting is that their skin can cling to this nicely if need be. However, when trying to cling they may find that the outer surface behaves in a different manner and is more slippery. Thankfully babies and children are very clever people who tend to learn and adapt to things like this pretty quickly. Even the clothes they are put in provide learning opportunities!

One layer on each

The next barrier level is a layer of clothing on each person, which blocks some of the sensory input for both. Depending on the materials, the clothing may be conducive to carrying or a hindrance. It's important to notice how different fabrics behave together, as their frictional properties against skin will be different to other materials. Do they produce a good amount of friction or does it make for little to no purchase?

So, what happens when the dyad needs to layer up when carrying? Well, they will have to work with what they've got, unless they're prepared to create a "carrying wardrobe". This is in no way essential for carrying, and most will have little inclination to overhaul their clothing. Having a look at the dyad's clothing is useful to see what materials work better together. The better the frictional properties of the fabric, the easier it will be for them to hold on as long as their movement isn't restricted, and the other fabric isn't a slippery one. It's unlikely that there won't be compatible materials and caregivers learn quickly what items are a hindrance to the carrying process, avoiding certain clothing when they know active carrying will be taking place.

Underwear

Underwear is also an interesting topic. Most babies here in the UK are nappied from birth. Even babies whose parents practice elimination communication tend to use nappies at least some, if not all of the time, especially in the early weeks and months. As we know, all clothing covering areas in contact with each body is essentially a barrier to carrying, but underwear is something pretty essential for most people, whether that's pants or nappies. The good thing is, many pants don't create a noticeable slipping effect due to the most common material for them being cotton. Many nappies are also ok in this respect too. The issues tend to come from bulk in this area, which changes the way in which a baby/child is able to hold on. This means we have to adapt how we carry based on what's on their bottom.

Disposable nappies tend to fold in on themselves but can be adjusted to create less of an issue by smoothing the bulk to the back. This does, however, tend to be a temporary measure as with movement the fold works its way back to the front. As well as a little bulk, it creates more of a gap and disconnect from the caregiver's body in this area. Fortunately, this doesn't usually present too much of a problem as babies tend to learn to adapt to this minor inconvenience.

The thinner, modern-style cloth nappies tend to behave in a way similar to disposables. Nappies which have been boosted to increase absorbency are the biggest barrier when it comes to underwear. It completely changes the way the body moves, how the body can make contact and puts distance between baby and caregiver. It's interesting that cloth is recommended to hold babies in an open position which is also sometimes recommended by experts to complement hip dysplasia treatment, but in an everyday sense it can hinder the carrying process. This is obviously a sore point to make, finding a fault with cloth, seeing as they are so much better for the environment than disposables, but I'm not saying they should be abandoned for disposable nappies, or that people need to buy thinner cloth. Just being aware of how each type of nappy behaves gives us more information to work with and gives us choices about whether or not we may want to use less boosting or give the baby a break from them from time to time to move in a natural way. It's interesting to note that babies tend to rarely eliminate on their caregiver without much warning beforehand and will usually signal repetitively and very strongly before giving up and doing so. To me this shows yet again how many amazing in-built human behaviours we have, how perfectly designed we are, and that so many things seem to be linked together to complement carrying and make it easier for us.

Socks and shoes

Other barriers to sensory input are socks and shoes. While shoes may be fairly obvious, it's often forgotten that socks and other flexible foot coverings are just another, suppler, variation of them. Socks tend to cling to the shape of the foot but restrict it in some ways, as well as being another barrier to skin. The potential implications of being shod much of the time has been interesting to think about. Babies wearing foot coverings are much less likely to initiate ankle and foot contact with the caregiver's body compared to when barefoot. If we look at the movements present in medial ankle and foot clinging, this is quite some work for the foot! It raises the question as to what implications for our feet not practicing this form of clinging has.

Also, of course, it can lessen the sensory input to trigger the reflexes of the foot, reducing chances of the reflexive action happening. It's incredible how much of an effect even the thinnest of foot coverings can have on this, resulting in many babies leaving their foot disconnected in a hip carry. Footed babygros tend to offer more freedom of movement for the foot and ankle in carrying, especially if it is oversized. Ones that fit well will have the length reduced as the leg lifts up and this both restricts movement and is actually physically painful over time (I know this first-hand from sleeping in an adult onesie which was the "perfect" fit height-wise), not to mention the fact it will leave the foot partially weightbearing.

As we discovered earlier in the book, the actions of the foot in clinging are specific to carrying during infancy. This movement is not happening in the same way elsewhere. Therefore, it is likely important that babies are given adequate opportunities to engage their feet in carrying so they may perform the functions they were designed for.

Sight

For the caregiver the barriers to sight are obviously mainly tied to where the baby/child is on their body. On their back they are unable to observe them and must rely on touch, sound and proprioception to work out how clinging is going. This can be an advantage though, as it means they can put more focus on these other sensory messages which sight sometimes interferes with. With the added limitation of how arms are able to move and reach behind the body, as well as the wider clinging surface in comparison to the hip, it's no wonder back clinging comes at a later stage in their development when their clinging capacity has expanded greatly, and the carrying person is relying less on sight. Back clinging tends to be more interesting for older children as they're more likely to be able to cling at a height where their head is positioned so they can see over the caregiver's shoulder. In younger babies who are physically able to cling to this different surface it tends to be for shorter periods and for more specific reasons (e.g. needing to free up the front of the body and both hands) and is less appealing to the child whose line of sight is blocked.

In chest-to-chest carrying the younger baby can also get frustrated from limited visual experience if they're at a state where they're craving interaction with the world. It's interesting that babies can sometimes go through stages where they are discontent in a caregiver-facing sling. Some of these babies will absolutely refuse facing in no matter what, but for others it may well be a case of not being afforded different positions on-body (whether in carrying or babywearing) that is contributing to their frustration. Another possibility is the caregiver expecting them to accept passivity (held in the carrier) at a time when they're in a state of alertness which is driving them to need interaction with the world.

Sound

Noisy environments and sound in general produce different barriers to carrying. Simply talking to someone other than the clinger leaves the carrying person less present, less able to "hear" other sensory communication. I'm not saying carrying should happen in silence, of course! We're going to experience sound around us pretty much all the time in different ways, and it's simply one of the things we get used to. We can acknowledge the ways in which sound may interfere in the carrying process for the individual caregiver, as well as for the baby/child. For example, in a very noisy environment we may find that they actually need to disengage from active carrying due to overload of one or more of the senses which can then interfere with other sensory input, meaning we switch to a passive hold.

Equilibrioception

Passive holds present barriers to equilibrioception, as the additional support provided by the caregiver takes away some or most of the body disruptions which come with active carrying. In effect, we can make them feel much more balanced in passive holds and their bodies are required to compensate less for our movements. If we use passive holds a lot when they're awake and active we're removing everyday opportunities for them to learn compensational movements in carrying and building an expectation into them that carrying requires little balance. Essentially, it's training them to rest.

PART V

Chapter 5.1

Further Thoughts

So, there you have it. How incredibly designed are we, as humans, to both carry our babies and to be active clingers? I hope you've found the information in this book as fascinating as I do, and that it has helped you to understand better how and why human babies are active clinging young. As interesting and plentiful as the available information is for certain things which are found in the carrying process, when it comes to piecing it all together to give a clear picture of all the ins and outs we're still very much in the theoretical stage. Whilst I've poured out just enough knowledge I have on the carrying process to make this book readable and manageable in size for the carrying enthusiast, it can still feel like the picture is blurry in places, because we don't always have concrete scientific evidence. At this point in time there are many theories to be explored about the carrying process. I've talked about some of these already, but in this penultimate chapter I'd like to introduce a few more and expand on some of what I've already discussed.

Attachment theory

Throughout this book we've had a primary focus on the way clinging works in a physical sense. As we head into the various theories, I'd like to begin by exploring the nature of attachment and the role active carrying plays in it. It's easy to forget to bring attachment into the clinging picture simply because we know that *holding* is an important touch-factor in forming secure attachment.[1] It's tempting to just focus on all the new areas of knowledge which comes with discovering clinging, but there are things to be observed in terms of potential attachment-forming activities which differ from passive carrying.

Secure attachment, as we know, relies on the prompt and appropriate response to babies' needs.[2] Evolutionary attachment theories line up with the physical evidence of an evolutionary expectation to have a primary caregiver. There's an advantage for the person whom which the baby has a primary secure attachment with to be the one who is also the main caregiver and carrying person. The one who is also securely attached to *you* is the one who responds to your needs in a timely manner and with sensitivity. The very nature of this links into what we've learned together about evolutionary advantages, showing us yet another way in which humans are designed to interact and form their relationship with a primary caregiver.

We've ascertained that clinging is vastly different to passive carrying, so what role may active carrying have in promoting a secure attachment? Well, clinging is yet another need to be met by the caregiver. Securely attached babies look to their caregiver/s to promote and support exploration of the varied environments and experiences they encounter. They also view people they're securely attached to as being a source of comfort and helper of emotional regulation. Knowing the sort of touch or carrying the baby specifically needs at the given time and providing the environment or opportunity to meet this need makes for a more attentive caregiver. The key is not simply holding by itself; it's correctly identifying and meeting that need. Now, I'm not saying that not meeting the need to cling prevents secure attachment – that would be absurd! We know that not meeting every single need for normal physical development doesn't in itself impact on attachment. It's erratic behaviours or non-responsiveness which does. I believe that meeting the need to cling can be compared to other protective factors in attachment: meeting specific needs.

The nature of clinging – movement, exercise, communication – and everything which goes with it benefits from a specific relationship with the caregiver. One which requires both parties to be present with each other, to listen and learn, and evolve as time goes on. It can obviously work with different forms of attachment, but a securely attached dyad works better together. It's been shown that increased holding may improve chances of the baby being securely attached, likely due to the physical connection causing increased responsiveness to the baby.[3] In that vein, it's not a big stretch to consider the potential impacts active carrying may also have. Hip clinging, for example, puts the child at an angle where it is very easy to make eye contact and communicate with each other in different ways to passive holds. This further facilitates responsiveness.

As some modern families come in different makeups to the evolutionary expected norm, again we're presented with the dilemma of "how do we adapt?". The focus of most research into attachment is focused on the gestational mother-baby dyad. I will reiterate; humans are a highly adaptable species. We know that forming their first attachment is not limited to either the mother or a person with a female body. Secure attachments can be made with any responsive and attentive caregiver, which shows how – even though there are biological advantages making the gestational parent primed to be the first attachment figure – babies are preadapted to survive by being born with the ability to form attachments with anyone who meets their needs. This leads me to believe that carrying is also adaptable, as it wouldn't make much sense if the person they formed the secure attachment to couldn't facilitate clinging. We know, for example, that male brains are rewired when they are the primary caregiver.[4] Are bodies affected by these changes too? It would be extremely beneficial to get answers to the question of whether non-gestational caregivers' bodies have ways to adapt to carrying just as brains adapt to parenting.

Much research has been conducted on the importance of human touch. We also know that holding babies affects them on a cellular level.[5] Babies need physical contact for proper attachment and development.[6] If the amount of holding a baby receives does this, how may active carrying affect both attachment and DNA further? The fact that human babies and children are designed to cling on-body must not be ignored. It's also interesting the effect that attachment can have on the physical health of the older child. A link has been discovered between secure attachment and increased levels of physical activity in adolescents.[7] Could this also effect the activity levels of the younger child or baby? Does a secure attachment provide additional motivation to interact in the world around a person?

Biological norm for primary carrying dyad

It consistently appears that female bodies – and, specifically, gestational parents – are inherently designed to be primary caregivers of babies and children. Whether or not this happens, due to many reasons, it seems clear from a biological point of view that this is what is expected of humans. As with so many other factors in parenting and life in general we aren't tied to this setup, and the adaptability of both babies and humans in general means we're able to modify carrying based on our family setup and needs. The important thing is that we recognise the biological baseline, acknowledge any benefits that go with it, and recognise that we're deviating from the biological norm when we do things differently. By doing this we're able to use this knowledge to adapt in ways which give our young the closest experience to what is biologically expected by them.

The multitude of genetic differences which favour compatibility of babies to female bodies makes this concept undeniable. They are the ones physically capable of pregnancy, birth and nourishment of the baby, and as we've seen, their bodies are designed in a way which makes carrying easier than for male bodies. This undoubtedly brings up questions surrounding what differences there may be in how babies develop their clinging behaviours if a female body isn't available for most of the carrying, and whether there are ways in which carrying can be adapted to better fit male bodies. As female bodies also come in many shapes and sizes it's likely this is possible.

It would be interesting to explore whether babies have physical adaptations specific to the gestational parent's body though.

In a rapidly changing and developing world, I wonder what we will discover going forwards. Evolution does not happen quickly. Rather than thinking that the way we were designed means that only gestational parents or those with female bodies should be primary caregivers, what it throws up for me is that maybe this gives even more weight to the belief that it takes a village to raise a child. So much can be adapted to modern times, yet there are still many ways of meeting the biologically expected needs. For example, where a person is unable to breastfeed, there's the option of a baby receiving breastmilk from another person, via breast or bottle. The biologically expected food is there if we source it. For carrying, if a person is physically unable to carry, there's more often than not another caregiver, relative or friend available daily or weekly to carry the baby. If the primary caregiver has a male body, again, there is the option for other female bodies to do some of the carrying. What I'm getting at here is that there are always ways of adapting to whatever situation we're in. The baby gets all of the other wonderful benefits of loving parents meeting their needs in specific ways, yet if there's something we know is needed but can't be provided in the expected way by one caregiver, it doesn't mean it's impossible to find a way of meeting the need.

The evolving future of civilisation is dependent on us taking a good hard look at current practices. This is across the board – how we're treating our planet, how we're treating ourselves and others – not just how we're parenting. There's such a disconnect going on, and a lot of it is down to attempts to "improve" life, free up more time, move less, connect with others in less meaningful ways, spend less time in nature. Our species doesn't need to turn its back on modern advances and differing traditions completely, but we certainly do need to be going back to basics and learning from our human design to inform us of what we truly need at the most basic level.

Normal clinging capabilities

I hypothesise that the clinging capabilities of the average human baby far surpasses what most of us believe to be possible. In my personal experience of mainly actively carrying one baby throughout his carrying years who showed no in-built physical advantages over other babies and being a person who carried him just as much as was needed (e.g. when he wanted to, no extra carrying, just observing carrying from our personal "normal"), I've seen that it doesn't take any extra work or commitment to develop these behaviours. As a young baby he was in the sling out of the house and it was only when he was around a year old that I began carrying him in-arms more for walks in the outdoors. Basically, I mainly used the resources available to me, followed his lead and haven't tried to shape his clinging development any further than adopting active carrying principles when he's been active and alert, and in the first year of his life still used a sling outside the house even when he was in this state. In fact, in terms of normal human behaviour, I've been pretty rubbish to be honest as I don't move anywhere near as much as is biologically expected of my body, especially since learning to drive before he was born. It also makes me wonder how his clinging would have developed had I only used the sling for when he needed rest.

Observing and working with other families who choose to adopt active carrying principles has shown me that it's not just a fluke. Even those who have needed to reteach their child how to cling have found it's usually simple enough to do when they've made a conscious effort to stick to active principles, rather than slipping back into passive ones. Babies are designed to cling and even when caregivers need to re-teach them to use their abilities it doesn't seem to be a mammoth task for most. Even my lower-tone child, Xander (3rd child), learned to cling long-past the time we switched from active to passive carrying, with much less carrying going on than with a younger baby. Clinging just looks and feels different with him – he needs supporting lower down his spine, so his legs don't do as much work as would be normal at his age (5.5 years old at the time of writing). I believe he would have a better clinging capacity if I hadn't switched to passive holds in his first year, and if I'd decided to carry him more when I started applying active principles to his and his older brother's carrying. Being pregnant then having a young baby meant I chose not to, yet he still reaped many benefits. I also wonder if he would have developed the sensory issues he has, or if they would at least be less pronounced, had he been through the normal clinging developmental process. Unfortunately, I will never know.

I also wonder if Logan (2nd child who was worn for long periods of time in the sling) has a biological advantage when it comes to clinging, seeing as he so easily learned to cling in carrying and he is almost 2 years older than Xander. He's always shown exceptional motor skills, kicking a football for hours each day from around a year old, mimicking specific fighting moves with precision from 2 etc. Since he learned to cling he's displayed behaviours such as scaling doorframes, and at 7 years old a random incident showed that he can cling independently on my hip with just his legs. I wonder how active carrying would have developed had he been encouraged to cling from the start. At this point in time I feel his spectrum of clinging potential is on the higher end.

Whilst I believe some aspects of advanced clinging behaviour rely on things such as leg and torso length, it appears that the easiest way to develop their inborn abilities are to work with them from the start. Thinking back to the critical period hypothesis, I do believe it's worthwhile to be aware of the windows of opportunity where it's easier to encourage voluntary clinging. The fact is, it doesn't cost a person anything to use active carrying principles. The hardest thing is reversing the conditioning their brain has had to automatically adopt passive principles. Even this doesn't always require too much effort once an awareness is present and a person commits to reminding themselves.

I'm very interested to see how Isaac's clinging behaviours develop once his carrying years pass. He likes to climb and swing from bars but doesn't get enough opportunities on a daily or even weekly basis. As parents we each prioritise different things and our individual circumstances tend to dictate what we focus on more. For me, I'm working towards us all moving more and spending more time outdoors by working on managing my health in better ways. As much as I'm a pioneer for active carrying, I'm not wanting it to be viewed or treated as something that is the "best". I believe in the idea of changing the basics of how we carry so that when we do it we're providing our children with what their bodies are designed to expect; to view clinging as simply normal. If that encourages us to move more, that's an added bonus.

So, it would be beneficial for us to gain a better understanding of what this "normal" looks like on a larger scale. This would require a high number of participants being taught active carrying principles, and a longitudinal study of how carrying progresses for each dyad. As a by-product of such a study it would be interesting to see how many babies experience retained reflexes and developmental delays, if any. Of course, this is me dreaming big, and while I would love this to happen I appreciate the enormity of such a task and the hurdles involved in getting research approved even for popular subjects. I like to think of what could be achieved in the future, though, and in taking steps such as writing this book I hope to plant seeds.

Weight in relation to physical development

What about heavier babies, though? I believe it's clear that clinging behaviours may be limited by excess weight of the baby/child in relation to their height, age and stage of physical development. Although weight in general has a negative impact on friction at a certain point, based on the type of materials against each other, weight in and of itself in carrying does not. Weight combined with a lack of clinging behaviours decreases friction, and as we know that clinging potential increases as physical development matures (which is also in line with growth and increased weight), it's clear that it's helpful for these to happen in unison. As we touched on earlier, motor development needs to occur in line with weight gain otherwise it is harder for the baby to stabilise their body, and delays may occur. It's been observed that overweight babies may be at greater risk of developmental delay.[8]

If the baby or child has no medical issues which affect their metabolism, movement or body make-up and they're given a level of freedom of movement which is biologically expected of them, then – along with a normal healthy diet – this should mean they will not be carrying excess weight. However, in a society where nutrition has both improved and deteriorated depending on how we feed ourselves, "normal" is still rather different today than in previous times. However, some babies will be on the upper end of the weight spectrum, even if exclusively breastfed, and we know that human milk is the biological norm for babies. Genetics will come into play for some babies, in that they have bigger built or taller parent/s. This would also be an interesting area to research, to find out whether these babies' clinging capacity is affected. It would also be worthwhile investigating clinging spectrums of lower weight babies, average ones and heavier to see how they may vary when all are carried using active carrying principles. The important thing though is to focus on developing the clinging behaviours rather than thinking they're too heavy to possibly ever be able to cling well enough. Working with the individual child's entry level clinging abilities and seeking to build them up by providing the relevant support to help them engage in the process without requiring what is beyond them is how they learn.

Reaching developmental milestones

Another interesting area for potential research is how active carrying may affect other areas of physical development. I believe we would see a shift towards reaching some milestones earlier than what is currently accepted as normal. As we saw, it's perfectly clear that the ways in which they use their bodies in on-body clinging are tied to other areas of physical development and that active carrying provides the practice and repetitive movements needed for them to grow stronger and develop normally. Babies are not required to cling with their legs in any other area of physical development at such a young age, though. Having regular experience of this action and movement may affect their development in ways currently unknown.

It's been noted that the increase in sedentary lifestyles has meant a change in the definition of "normal" ages for achieving developmental milestones.[9] We know the importance of movement for normal physical development and excessive "containerisation" of the infant is consistently referred to as a risk factor for the retention of reflexes. Retained reflexes then delay motor development. Whilst not all children will retain reflexes, the delays in carrying out the required repetitions of movements can delay integration, and this offers insight into possible reasons for a trend of milestones being reached later over the years. What may happen if some time where a baby would otherwise be sedentary was replaced with active carrying? This could be as easy as changing support in regular carrying situations. It would be wonderful if research was conducted in this area to see how clinging vs. passive containment affects development. Is on-body clinging a "missing link" in their physical development?

Caregivers' bodies

Something which occupies my thoughts is how caregiver's bodies are impacted by normal, aligned carrying. Of course, it's easier to notice how abnormal and unaligned carrying would impact on the body, because it's more visual and complaints are more likely to occur from it, alerting us to the negatives. The positives can be harder to quantify. We do know that loadbearing improves bone mineral density[10] and its needed for good joint and bone health. If human babies are designed to cling, and their bodies are born expecting it, what does that mean for our bodies? Are gestational parents' bodies biologically expecting to carry actively? How does this impact on our health later on in life? The fact that bone mineral density reduces during pregnancy and lactation[11] could signify that the loadbearing activity of carrying may be nature's way of helping to recoup it postpartum, alongside diet.

A potential study could focus on one group who are taught active carrying principles and shown how to do so in an aligned manner, with a control group of caregivers carrying in what is considered "normal" ways (e.g. passive holds, likely misaligned to some extent). There is research showing that using a sling on one side of the body can increase risk of scoliosis[12] so it seems reasonable to postulate that passive carrying out of alignment may contribute to issues of the spine. Also, being able to monitor for any benefits to the musculoskeletal system would be advantageous.

Another thought about caregivers' bodies is the potential impact carrying may have on the exercise element of it. Does a clinging child encourage carrying more often? Is a natural consequence of having a clinger a knock-on effect of relying on "containers" less? For example, using carrying as a form of entertainment and interaction with the world as opposed to, say, a baby walker. Would it impact on the caregiver's choices – and need for outsourcing their arms – if carrying was an easy and pleasurable experience for them? From observations and my own experience, I've seen a definite trend towards defaulting to arms when the baby is able to cling.

Active carrying as a pleasurable experience

I've heard some comments about whether we're expecting too much of the child to learn to cling – is it a chore for them? One of these incidences was to do with someone's husband noting how much harder it was for the baby to cling to him than his mother. This to me shows again how there is a learning curve for the baby (and caregiver!) when being asked to cling to a new environment, especially when that one is shaped differently. It isn't necessarily a negative experience for a baby to learn to adapt to a different body, but it's important that both caregiver and child are happy and comfortable with the learning experience. Another comment was to do with a preconceived idea of the baby not being able to learn such a thing because the mother had carried passively for so long and initial attempts at engaging her baby in clinging appeared to not be working. She felt that he simply did not have the capacity to do so and that trying to "force" him to do so was both wrong and futile. With further discussion we identified that she was supporting in a place unsuitable for his clinging capacity and once she changed this he displayed a better ability to cling. It's important to recognise that yes, there are ways in which we may inadvertently make the learning experience less pleasurable, but that it is not inherently so.

An interesting discovery was made by Zelazo et al. in 1994.[13] It was in relation to a previous study the year before, which we looked at in chapter 1.3. They looked back over the videos of babies performing the reflexive stepping action and practicing sitting to measure the incidence of smiling during this time. The theory was that associating an experience with achievement brings pleasure, and that for this to happen, a period of time must pass for them to effectively create memories of previous times of performing the action. They found that as time went on the babies smiled most of the time when participating in the exercise.

Therefore, it's not too big a leap to ponder if active carrying is a pleasurable – rather than a necessary – experience for the baby. It's an experience of repetition, learning, getting more capable and refined at it, and we know that babies and children thrive on learning new things. The clinging baby engages with both the caregiver and the environment around them, which meets a need for connection to them as well as learning and engaging in the environment around them. These in and of themselves are pleasurable experiences. If active carrying was a negative experience we would expect to see some form of protestation and/or disengagement, as in other activities. In fact, we do see these behaviours when a baby or child is fatigued, therefore this in itself lends credence to the idea that clinging is not inherently uncomfortable or unenjoyable but may become so when they tire. It would be fascinating to explore the ways in which active carrying affects the pleasure of the baby.

Energetic Costs of Carrying

We're also told that slings and carriers were invented due to the increased locomotor costs of walking upright; that carrying babies in-arms was too much effort for us, so we needed to invent something to conserve our energy. You know where I'm going with this, don't you? This theory was created based on not understanding the inborn clinging abilities – and the capacity to learn how to cling harder - of human babies. A clinging baby is a lighter baby – lighter than one in a sling. When they are in a sling/carrier they are a dead weight, to a certain degree (with the full force of this being when asleep). Again, the premise of evolution is improvement – change for the better. Why "enhance" one area whilst leaving our young at a disadvantage, and actually in danger, by removing this hair that is so great for holding onto, which in turn would assist clinging?

Clinging works to reduce the energetic costs of carrying babies and children. A baby/child who clings is doing much of the work involved, and their perceived weight is less than a static, non-engaged load. This is important to acknowledge as current popular belief is that in-arms carrying is energy-expensive, especially in asymmetrical positions, and that securing a child to the trunk (babywearing) is much more cost-effective.[14] This study tends to be cited in the babywearing industry by some to promote another "benefit" of babywearing, and this is one of the ways in which in-arms myths are perpetuated. It can come across that in-arms carrying is tiring and babywearing is easier to do. With the majority holding their babies in passive ways, it's easy to see why this research would be easily believed and used as a promotion of slings and carriers.

It's worth being aware that this study was conducted with inactive weights, and a weighted mannequin was used for the hip carrying test. Obviously, babies and children are more active than a dead weight even in passive holds, so that in itself muddies the findings. The study did not explore active carrying/clinging. Furthermore, a picture of the in-arms test shows the subject walking on the treadmill with hip thrust out, with the mannequin's legs dangling straight down and its upper body leaning backwards at a near-45° angle. All of these things will contribute to additional energetic costs. The researchers did acknowledge the fact the mannequin can't compare to a real child's activity in carrying but concluded that a mannequin doesn't wriggle and that it is easier to carry it in the most comfortable position for the carrying person.

It's important for research to be conducted using real babies and children, as the researchers conceded. However, active carrying/clinging must be included in this rather than just focusing on the passive carrying/holding, which is how most people carry. I believe the findings of such a study would be very different and would shine a light on one of the many benefits of active carrying.

Chapter 5.2

What We Need from the Scientific and Wider Community

As you can see I have many thoughts on the carrying process, yet these are still just a small selection. There are many things to consider, other avenues to explore, and much discussion to be had. Many of the things we observe about in-arms seem to give overwhelming anecdotal evidence that carrying clinging is an activity expected of the human baby. However, to have clinging recognised by the scientific community as a normal infant behaviour and recognised as a developmental process in its own right we will need more investigation and clinging-specific research conducted. Going forwards, it's clear that we need much input from the scientific community and specialists in fields such as occupational therapy, osteopathy, biomechanics and so forth. It is imperative that clinging be recognised for what it is – a developmental process. It is abundantly clear that the development of clinging follows and complements other motor development. Once it is recognised as a normal human behaviour I believe the potential for our understanding will grow exponentially.

As active carrying gains more attention around the world, increasing numbers of caregivers are seeing their children's clinging capacities develop and grow when applying active carrying principles. The concept is very much in its infancy though, and one way of confirming what we already know, as well as to discover the vast amounts that we don't, is for subject-specific research to be conducted. We also need specialists bringing their deep knowledge to individual parts of the developmental process and a coming together of minds to take this subject forward. Collaboration benefits us all.

Going forwards, it would be useful to have studies looking at things like:

- What our current "normal" is for clinging behaviours and how it changes when a child has completed the developmental process of active carrying

- Identifying in greater detail the effects that barriers have to sensory input in carrying

- The differences in how babies interact with caregivers during active and passive carrying

- In what ways completing the clinging developmental process impacts on health and strength of the body after carrying ceases

- The extra functions of vellus hair

- The ways in which clinging – and especially skin-to-skin clinging - affects the development of the nervous system

- Whether completing this developmental process has an impact on the timescale for reaching other milestones

- Whether active carrying can be used therapeutically (e.g. helping integrate reflexes or aid with non-invasive DDH treatment)

From specialists, I would love to see occupational therapists adding to the reflexes knowledge by offering insights into their take on the carrying reflexes, and whether carrying could potentially be a therapeutic aid or be put forward as a possible preventative measure for the risk of children developing learning difficulties. If parents are encouraged to work naturally with the reflexes of their babies, would we eventually see a fall in the retention of some reflexes? People with greater knowledge of reflexes may also be able to see the carrying process in a different light and put forward their ideas of how each one affects it. Knowledge of encouraging integration in the older baby or child with retained reflexes may help evolve the teaching of clinging to one who hasn't followed the normal developmental process of carrying. Work performed with ones on the autistic spectrum, and with other conditions such as cerebral palsy, may shine a light on how to facilitate carrying children with barriers to clinging.

From chiropractors and osteopaths, it would be great to see carrying-specific treatment plans for caregivers which include guidance as to how best to support the baby. At the moment the focus appears to be on correcting passive holding techniques but having knowledge of active carrying to pass on would be beneficial.

From movement specialists it would be amazing to get their input on how clinging influences the ways in which we move – and are *able* to move – as we outgrow on-body clinging. Also, how carrying in various ways impact on the caregiver's body. It would be good to explore further the ways in which babies' clinging is impacted by the caregiver's posture and alignment, their gait and the surfaces they walk upon. Just from actively carrying a baby over different terrains it's apparent that they behave differently. Getting the deeper biomechanical insights into things like this would both improve knowledge and pave the way for enhancing family movement classes.

For the babywearing industry, I would love to see an understanding of active carrying and the differences between it, holding and babywearing. At a time where there is a movement to change the terminology surrounding slings and carriers because of a growing aversion to the term "babywearing", the true concept of in-arms is lost. There are huge differences between clinging and babywearing. There are differences between clinging and holding, as well as holding and babywearing. Terminology to define different ways of carrying children are needed so that the benefits of each activity are not effectively erased. Carrying is an umbrella term. At this point in time it is harmful to use it as a replacement for the word babywearing. Babywearing educators are in a position of influence and the language a person uses has power.

From educators, I would love to see the integration of in-arms knowledge in babywearing teaching. In-arms holds many keys to developing babywearing expertise. It's massively influenced my personal and professional knowledge and helped me to revolutionise my babywearing consultancy course through Carried. Caregivers learn about all aspects of parenting through many different ways, and through babywearing there is a great opportunity for them to learn more about in-arms carrying. As it's such an ancient concept it's little wonder there's such a lack of knowledge about clinging in the general public and even in specialist areas. This can change, and this past year has shown the domino effect of education being absorbed then passed on, creating new awareness and exploration.

For the babywearing industry itself, in-arms knowledge can help manufacturers with product development. Slings and carriers cannot replace in-arms carrying, and we should not be seeking to, but things such as positioning, and support can offer clues to making Western babywearing better. Buckled carrier design especially can be improved with in-arms understanding.

It's also incredibly important to learn from caregivers themselves. Individual carrying experiences will vary greatly and we can learn much from understanding the many variations development brings. Caregivers spend the most time with the child and know them the best. The more dyads practicing active carrying, the more we can learn. By giving them information to then make an informed decision about active carrying we provide them with an opportunity they may not have had otherwise. Some of my ideas of points of interest going forwards are:

- Looking at active carrying on male bodies – my professional experience in this area has been limited, so it would be interesting to find out more about how different babies and children adapt to a less-adapted clinging environment. Also, how does clinging develop when the baby is being brought up solely by one or more males?

- Exploring the nature of clinging in terms of gestational caregiver vs. non-gestational (NG), where the latter has a female body – are gestational bodies adapted to the babies they give birth to (or vice versa) and is clinging any harder for babies being carried by NG bodies?

I hope the contents of this book provide a launchpad for taking the exploration and understanding of in-arms further.

References

PART I

Chapter 1.1 – The Clinging Young Concept

[1] Tierjunges und Menschenkind im Blick der vergleichenden Verhaltensforschung –Hassenstein, B (1970)

[2] Das Tragen des Säuglings im Hüftsitz – eine speziell Anpassung des menschlichen Traglings – Kirkilionis, E (1992)

[3] Examples include: Wikipedia - https://en.wikipedia.org/wiki/Baby_transport

Chapter 1.2 – An Introduction to How Clinging Works

[1] The Primitive Reflexes: Considerations in the Infant – Berne, SA (2006)

[2] The Critical Period - Sengpiel, F (2007)

[3] Are there critical periods for musical development? Trainor, LJ (2005)

[4] Critical periods in human growth and their relationship to diseases of aging – Cameron, N; Demerath, EW (2002)

[5] A Baby Wants to be Carried – E Kirkilionis

[6] Term used by biomechanist, Katy Bowman

[7] Infant calming responses during maternal carrying in humans and mice - Esposito, G; Yoshida, S; Ohnishi, R; Tsuneoka, Y (2013)

[8] Influences of mechanical stress on prenatal and postnatal skeletal development – Carter, DR; Orr, TE; Fyhrie, DP; Schurman, DJ (1987)

[9] See Chapter 1.1, reference 3.

[10] The Containerization of Infants – Breitbach, B (2010)

[11] Postural development in school children: a cross-sectional study - Lafond, D; Descarreaux, M; Normand, MC; Harrison, DE

[12] Does physical activity attenuate, or even eliminate, the detrimental association of sitting time with mortality? A harmonised meta-analysis of data from more than 1 million men and women - Ekelund, U; Steene-Johannessen, J; Brown, WJ (2016)

[13] Vitamin "Flat and Level" OVERDOSE – Katy Bowman (Web article, January 16, 2015)

Chapter 1.3 – The Role of Reflexes

[1] The Rhythmic Movement Method: A Revolutionary Approach to Improved Health and Well-Being – Blomberg, H

[2] Conditioned reflexes: An investigation of the physiological activity of the cerebral cortex - Pavlov, PI (1927)

[3] Reflexes and Their Relationship to Behavioural State in the Newborn - Lenard, HG; von Bernuth, H; Prechtl, HFR (1968)

[4] The impact of institutionalization on child development – Maclean, K (2003)

[5] Structural foundations of behaviour – McGraw, MB (1943)

[6] A longitudinal study of the Babinski and plantar grasp reflexes in infancy - Dietrich, HF (1957)

[7] "Walking" in the Newborn - Zelazo, PR; Zelazo, NA; Kolb, S (1972)

[8] A dynamic systems approach to the development of cognition and action - E Thelen, LB Smith

[9] Infant Motor Development – Piek, JP

[10] Factors influencing the asymmetrical tonic neck reflex in normal infants - Coryell, J; Henderson, A; Liederman, J (1982)

[11] Prone and Supine Positioning Effects on Energy Expenditure and Behavior of Low Birth Weight Neonates - Masterson, J; Zucker, C; Schulze, K (1987)

[12] Locomotor Primitives in Newborn Babies and Their Development – Dominici, N et al. (2011)

[13] Normal Development and Deviations in Development of the Nervous System - Samuels, MA.; Ropper, AH (2009)

[14] See reference 1

[15] Reflex mechanisms in the development of prehension – Twitchell, TE (1970)

[16] Skin-to-skin contact (kangaroo care) promotes self-regulation in premature infants: Sleep-wake cyclicity, arousal modulation, and sustained exploration - Feldman, R; Weller, A; Sirota, L; Eidelman, AI (2002)

[17] Repetitive processes in child development – Bower, TGR (1976)
    From reflexive to instrumental behaviour – Zelazo, P (1976)

[18] Specificity of practice effects on elementary neuromotor patterns – Zelazo, NA; Zelazo, P; Cohen, KM; Zelazo, PD (1993)

[19] Clinical significance of plantar grasp response in infants - Futagi, Y; Suzuki, Y; Goto, M (1999)

[20] Longitudinal motor development of "apparently normal" high-risk infants at 18 months, 3 and 5 years - Goyen, TA; Lui, K (2002)

[21] Infant Stimulation: Modification of an Intervention Based on Physiologic and Behavioral Cues – Burns, K; Cunningham, N; White-Traut, R; Silvestri, J; Nelson, MN (1994)

Chapter 1.4 – Breastfeeding and Carrying

[1] Co-evolution of lactation and suckling illuminates the adaptive challenges of prematurely born infants – Jeffrey Alberts (Ultra-early Intervention Conference, Stockholm – 2018)

[2] Breast crawl research

[3] Care of the parents in 'Care of the high-risk neonate' – Klaus, MH; Kennel, JH (2001)

[4] Uterine massage for preventing postpartum haemorrhage - Hofmeyr, GJ; Abdel-Aleem, H; Abdel-Aleem, MA (2008)

[5] Optimal positions for the release of primitive neonatal reflexes stimulating breastfeeding - Suzanne D.Colson, Judith H.Meek, Jane M.Hawdon (2008)

[6] Use of baby carriers to increase breastfeeding duration among term infants: the effects of an educational intervention in Italy - Alfredo Pisacane, Paola Continisio, Cristina Filosa, Valeria Tagliamonte, Grazia Isabella Continisio (2012)

PART II

Chapter 2.1 – Participation in Carrying - Where Does It begin?

[1] Anticipatory adjustments to being picked up in infancy - Reddy, V; Markova, G; Wallot, S (2013)

[2] Postural adjustments due to external perturbations during sitting in one-month-old infants: evidence for the innate origin of direction specificity – Hedberg, Å; Forssberg, H; Hadders-Algra, M (2004)

[3] Dunstan Baby Language - https://en.wikipedia.org/wiki/Dunstan_Baby_Language
　　DBL research - www.dunstanbaby.com/our-research/

[4] Does Infant Carrying Promote Attachment? An Experimental Study of the Effects of Increased Physical Contact on the Development of Attachment - Anisfeld, E; Casper, V; Nozyce, M; Cunningham, N (1990)

[5] The emergence of language: On being picked up – Lock, A (1984)

[6] Intonation and communicative intent in mothers' speech to infants: Is the melody the message? – Fernald, A (1989)

[7] Autistic disturbances of affective contact – Kanner, L (1943)

[8] Eshkol–Wachman movement notation in diagnosis: The early detection of Asperger's syndrome - Teitelbaum, O; Benton, T; Shah, PK; Prince, A; Kelly, JL; Teitelbaum, P (2004)

Chapter 2.2 – Developmental Process

[1] See Chapter 1.2, reference 8

[2] Towards an understanding of stereotypic behaviour in laboratory macaques – Philbin, N

[3] Neonatal behavioral assessment scale - Brazelton, TB;, Nugent, JK; Lester, BM (1987)

[4] "Neonate" definition - Merriam-Webster online dictionary

[5] Des menschliche Säuling als Tragling unter Besonderer Berücksichtigung der Prophylaxe gegen Hüftdysplasie – Kirkilionis, E (1989)

[6] "Landau reflex" definition – Medical Dictionary, TheFreeDictionary (online dictionary)

[7] Birth-to-5 development timeline – NHS UK

[8] WHO growth standards (2006)

Chapter 2.3 – Communication in Carrying

[1] Natural Movement and Eyes, Podcast Ep. 45 – Nutritious Movement (March 1, 2016)

[2] Gaze following in newborns - Farroni, T; Massaccesi, S; Pividori, D; Johnson, MH (2004)

PART III

Chapter 3.1 – Physiology and Anatomy of Babies

[1] Motor development during infancy and early childhood: Overview and suggested directions for research – Malina, RM (2004)

[2] Healthy posture for babies and children: Tools for helping children to sit, stand and walk naturally – Kathleen Porter

[3] Primitive reflexes and Attention-Deficit/Hyperactivity Disorder: Developmental origins of classroom dysfunction - Taylor, M; Houghton, S; Chapman, E (2004)

[4] Primitive reflexes and postural reactions in the neurodevelopmental examination -Zafeiriou, DI (2004)

[5] See Chapter 1.3, reference 1

[6] See reference 2

[7] See Chapter 1.2, reference 6

[8] See Chapter 1.2, reference 4

[9] See Chapter 2.2, reference 5

[10] Should neonates sleep alone? – Morgan, BE; Horn, AR; Bergman, NJ (2011)

[11] Development of the osseous and cartilaginous acetabular index in normal children and those with developmental dysplasia of the hip (A cross-sectional study using MRI) – Li, LY; Zhang, LJ; Li, QW; Zhao, Q; Jia, JY; Huang, T (2012)

[12] Ätiologie, prophylaxe und frühbehandlung der luxationshüfte – Büschelberger, H (1964)

[13] Carrying babies or toddlers in baby carriers or shawls – Dr. E. Fettweis

[14] Das tragen des säuglings im hüftsitz – Kirkilionis, E (1992)

[15] Normal ranges of hip motion in infants six weeks, three months and six months of age - Bleck, EE (1975)

[16] Causes of DDH – IHDI

[17] Developmental dysplasia of the hip – NHS

[18] See reference 17

[19] See reference 16

[20] See reference 16

[21] Educational statement – IHDI

[22] See reference 15

[23] See chapter 1.2, reference 4

[24] Body fat in neonates and young infants: validation of skinfold thickness versus dual-energy X-ray absorptiometry - Schmelzle, HR; Fusch, C (2002)

[25] The expensive-tissue hypothesis: the brain and the digestive system in human and primate evolution - Aiello, LC; Wheeler, P (1995)

[26] Body fat activity levels - Relation of activity levels to body fat in infants 6 to 12 months of age - Li, R; O'connor, L; Buckley, D; Specker, B (1995)

[27] Bone growth in length and width: The yin and yang of bone stability - F. Rauch (2005)

[28] Biomechanics – London College of Osteopathy and Health Sciences

[29] "Shoes" Practice of pediatric orthopaedics – Staheli, LT

Chapter 3.2 – Differences Between Caregivers' Bodies

[1] Variability in anatomical features of human clavicle: Its forensic anthropological and clinical significance – Sehrawata, JS; Pathak, RK (2016)

[2] Directional asymmetry in the human clavicle - Mays, S; Steel, J; Ford, M (1999)

[3] Postnatal growth of the clavicle: birth to 18 years of age - McGraw, MA; Mehlman, CT; Lindsell, CJ; Kirby, CL (2009)

[4] Developmental evidence for obstetric adaptation of the human female pelvis - Huseynov, A; Zollikofer, CPE; Coudyzer, W; Gascho, D; Kellenberger, C; Hinzpeter, R; Ponce de Leon, ME (2016)

[5] Pelvic Breadth and Locomotor Kinematics in Human Evolution - Gruss, LT; Gruss, R; Schmitt, D (2017)

[6] Study of the carrying angle of the human elbow joint in full extension: a morphometric analysis - Paraskevas, G;

Papadopoulos, A; Papaziogas, B; Spanidou, S; Argiriadou, H; Gigis, J (2004)

[7] Gender differences in strength and muscle fiber characteristics – Miller, AEJ; MacDougall, JD; Tarnopolsky, MA; Sale, DG (1992)

[8] Healthy percentage body fat ranges: an approach for developing guidelines based on body mass index - Gallagher, D; Heymsfield, SB; Heo, M; Jebb, SA; Murgatroyd, PR; Sakamoto, Y (2000)

[9] How much weight will I gain during pregnancy? - NHS

[10] Chest skin temperature of mothers of term and preterm infants is higher than that of men and women - Bauer, K; Pasel, K; Versmold, H (1996)

[11] See chapter 1.4, reference 4

[12] Definition of the problem - The evolution of human reproduction - Short, RV (1976)

Chapter 3.3 – Side Preferences

[1] Can the archaeology manual of specialization tell us anything about language evolution? A survey of the state of play – James Steel, Natalie Uomini (2009)

[2] Why are more people right-handed? - Holder, M. K. (1997)

[3] Neuropsychology: The Neural Bases of Mental Function - Banich, Marie (1997)

[4] Left-side preference for holding and carrying newborn infants. Parental holding and carrying during the first week of life - de Château, P (1983)

[5] A prospective study of the development of laterality: neonatal laterality in relation to perinatal factors and maternal behavior – Thompson, AM; Smart, JL (1993)

[6] Infant holding biases and their relations to hemispheric specializations for perceiving facial emotions – Vauclair, Jacques; Donnot, Julien (2005)

[7] The left-side holding preference is not universal: Evidence from field observations in Madagascar - Masayuki Nakamichi (1996)

[8] Left-side preference for holding and carrying newborn infants. II: Doll-holding and carrying from 2 to 16 years - de Château, P; Andersson, Y (1976)

[9] Unimanual to bimanual: Tracking the development of handedness from 6 to 24 months - Nelson, EL; Campbell, JM; Michel, GF (2013)

[10] Ask a grown-up: why do we have two breasts when we need only one? – Pixie McKenna

PART IV

Chapter 4.1 – Evolution and Carrying

[1] Evolution of nakedness in Homosapiens – Rantala, MJ (2007)

[2] Origin of bipedalism – Anthropology World

[3] Timeline of human evolution – Wikipedia

[4] Ontogeny of locomotion in mountain gorillas and chimpanzees – Doran, DM (1997)

[5] The energetic cost of locomotion: humans and primates compared to generalized endotherms - Steudel-Numbers, KL (2003)

[6] A nearly complete foot from Dikika, Ethiopia and its implications for the ontogeny and function of Australopithecus

afarensis - DeSilva, JM; Gill, CM; Prang, TC; Bredella, MA; Alemseged, Z (2018)

[7] Measurement of selected hip, knee, and ankle joint motions in newborns - Gilmore-Waugh, K; Minkel, JL; Parker, R; Coon, VA (1983)
Normal range of motion of joints in male subjects - Boone, DC; Azen, SP (1979)

[8] Hip extensor mechanics and the evolution of walking and climbing capabilities in humans, apes, and fossil hominins - Kozma, EE; Webb, NM; Harcourt-Smith, WEH; Raichlen, DA; D'Août, K; Brown, MH; Finestone, EM; Ross, SR; Aerts, P; Pontzer, H (2018)

[9] Allometry of primate hair density and the evolution of human hairlessness - Schwartz, GG; Rosenblum, LA (1981)

[10] The biology of hair – Ebling, FJ (1987)

[11] The great arc of dispersal of modern humans: Africa to Australia – Oppenheimer, S (2009)

[12] Vitamin D: in the evolution of human skin colour - Yuen, AWC; Jablonski, NG (2010)

[13] See reference 1

[14] The Human Career: Human Biological and Cultural Origins – R.G. Klein (2009)

[15] See reference 1

[16] Genetic analysis of lice supports direct contact between modern and archaic humans – Reed, DL; Smith, VS; Hammond, SL; Rogers, AR; Clayton, DH (2004)

[17] Genetic variation at the MC1R locus and the time since loss of human body hair - Rogers, AR; Iltis, D; Wooding, S (2004)

[18] Mechanical analysis of infant carrying in hominoids – Amaral, LQ (2008)

Chapter 4.2 – Frictional Properties of Skin and Hair

[1] Friction - Wikipedia

[2] Coefficient of friction – Wikipedia

[3] The biomechanical properties of the skin - Hussain, SH; Limthongkul,B; Humphreys, TR (2013)

[4] Maternal nutrition and the regulation of milk synthesis - Hartmann, P; Sherriff, J; Kent, J (1995)

[5] Friction and lubrication of human skin - Adams, MJ; Briscoe, BJ; Johnson, SA (2007)

[6] Contribution of stratum corneum in determining bio-tribological properties of the human skin - Pailler-Mattei, C.; Pavan, S.; Vargiolu, R.; Pirot, F.; Falson, F.; Zahouani, H. (2007)

[7] Tribology of skin: review and analysis of experimental results for the friction coefficient of human skin - Derler, S; Gerhardt, LC (2012)

[8] On the role of adhesive forces in the tribo-mechanical performance of ex vivo human skin – Morales-Hurtado, M.; de Vries, E.G.; Peppelman, M.; Zeng, X.; van Erp, P.E.J.; van der Heide, E. (2017)

[9] Frictional properties of human skin: relation to age, sex and anatomical region, stratum corneum hydration and transepidermal water loss – Cua, AB; Wilhelm, KP; Maibach; HI (1990)

[10] In vitro sun protection factor determination of herbal oils used in cosmetics - Kaur, CD; Swarnlata, S (2010)

[11] Regulation of human skin pigmentation and responses to ultraviolet radiation - Miyamura, Y et al. (2006)

[12] Cancer Council (Australia)

[13] See reference 9

[14] Cohesion (chemistry) – Wikipedia

[15] Skin physiology of the neonate and young infant: A prospective study of functional skin parameters during early infancy - Hoeger, PH; Enzmann, CC (2002)

[16] Sebum levels during the first year of life - Agache, P; Blanc, D; Barrand, C; Laurent, R (1980)

[17] Why do we have apocrine and sebaceous glands? - Porter, AMW (2001)

[18] Effect of aging on sebaceous gland activity and on the fatty acid composition of wax esters - Yamamoto, A; Serizawa, S; Ito, M; Sato, Y (1987)

[19] Embryology of hair in the biology of hair growth – Pinkus, F (1958)

[20] The developing human: clinically oriented embryology - Moore, KL (2011)

[21] Biotechnology in hair care (I): overview - Kessler-Becker, D (2016)

[22] Physiology of the vellus hair follicle: hair growth and sebum excretion - Blume, U; Ferracin, J; Verschoore, M; Czernielewski, JM; Schaeffer, H. (1991)

[23] Hairless mutation: a driving force of humanization from a human–ape common ancestor by enforcing upright walking while holding a baby with both hands - Sutou, S (2012)

[24] Avoidance of overheating and selection for both hair loss and bipedality in hominins – Ruxton, GD; Wilkinson, DM (2011)

[25] A practical model for hair mutual interactions - Chang, JT; Jin, J; Yu, Y (2002)

Chapter 4.3 – Senses in Carrying

[1] Postnatal development of vision in human and nonhuman primates - Boothe, RG; Dobson, V; Teller, DY (1985)

[2] Neonatal recognition of the mother's face - Bushneil, IWR; Sai, F; Mullin, JT (1989)

[3] Cardiac responses on the visual cliff in prelocomotor human infants - Campos, JJ; Langer, A; Krowitz, A (1970)

[4] Binocular vision and spatial perception in 4- and 5-month-old infants – Granrud, CE (1986)

[5] GIF shows how baby's eyesight develops over 12 months – Clinic Compare & Moorfields Eye Hospital

[6] Emotions and emotional communication in infants - Tronick, EZ (1989)

[7] The role of eye-to-eye contact in maternal-infant attachment - Robson, KS (1967)

[8] Olfaction in the development of social preferences in the human neonate - Macfarlane, A (1975)

[9] A possible protocognitive role for odor in human infant development - Vantoller, S; Kendalreed, M (1995)

[10] The Oxytocin Factor – Uvnas-Moberg, K

[11] Development of hearing. Part III. Postnatal development – Peck, JE (1995)

[12] Behavior during the prenatal period: Adaptive for development and survival - Hepper, P (2015)

[13] Morphology of cutaneous receptors - Iggo, A; Andres, KH (1982)

[14] Emotion regulation via maternal touch - Hertenstein, MJ; Campos, JJ (2001)

[15] The communication of emotion via touch - Hertenstein, MJ; Holmes, R; McCullough, M; Keltner, D (2009)

[16] Epigenetic correlates of neonatal contact in humans - Mah, SM; Barr, RG; Boyce, WT; Kobor, MS (2017)

[17] The effect of skin-to-skin contact (kangaroo care) shortly after birth on the neurobehavioral responses of the term newborn: a randomized, controlled trial - Ferber, SG; Makhoul, IR (2004)

[18] The Vital Touch: How Intimate Contact with Your Baby Leads to Happier, Healthier Development - Heller, S.

[19] Frequency of infant stroking reported by mothers moderates the effect of prenatal depression on infant behavioural and physiological outcomes - Sharp, H; Pickles, A; Meaney, M; Marshall, K; Tibu, F; Hill, J (2012)

[20] The Power of Touch – Chillot, R (2013)

PART V

Chapter 5.1 – Further Thoughts

[1] Infancy in Uganda: Infant care and the growth of love - Ainsworth, MDS (1967)

[2] Attachment theory – Bowlby, J

[3] See chapter 2.1, reference 4

[4] Father's brain is sensitive to childcare experiences - Abraham, E; Hendler, T; Shapira-Lichter, I; Kanat-Maymon, Y; Zagoory-Sharon, O; Feldman, R (2014)

[5] See chapter 4.3, reference 16

[6] The Importance of Touch in the Development of Attachment - Duhn, L (2010)

[7] Attachment relationships and physical activity in adolescents: The mediation role of physical self-concept - Li, R; Bunke, S; Psouni, E (2016)

[8] Infant overweight is associated with delayed motor development - Slining, M; Adair, LS; Davis Goldman, B; Borja, JB; Bentley, M (2010)

[9] See chapter 1.2, reference 8

[10] The effect of weight-bearing exercise on bone mineral density: A study of female ex-elite athletes and the general population – Etherington, J; Harris, PA; Nandra, D; Hart, DJ; Wolman, RL; Doyle, DV; Spector, TD (2009)

[11] The effects of pregnancy and lactation on bone mineral density – More, C; Bettembuk, P; Bhattoa, HP; Balogh, A (2001)

[12] The effects of body posture by using baby carrier in different ways - Kim, K; Yun, KH (2013)

[13] Infant smiling during stimulation of stepping and sitting - Zelazo, NA; Zelazo, PR; Cohen, K; Zelazo, PD (1994)

[14] The energetic costs of load-carrying and the evolution of bipedalism - Watson, J.C.; Payne, R.C.; Chamberlain, A.T.; Jones, R.K.; Sellers, W.I. (2008)

www.ingramcontent.com/pod-product-compliance
Lightning Source LLC
Chambersburg PA
CBHW021352210526
45463CB00001B/83